Centerville Library
Washington-Centerville Public Library
Centerville, Ohio
DISCARD

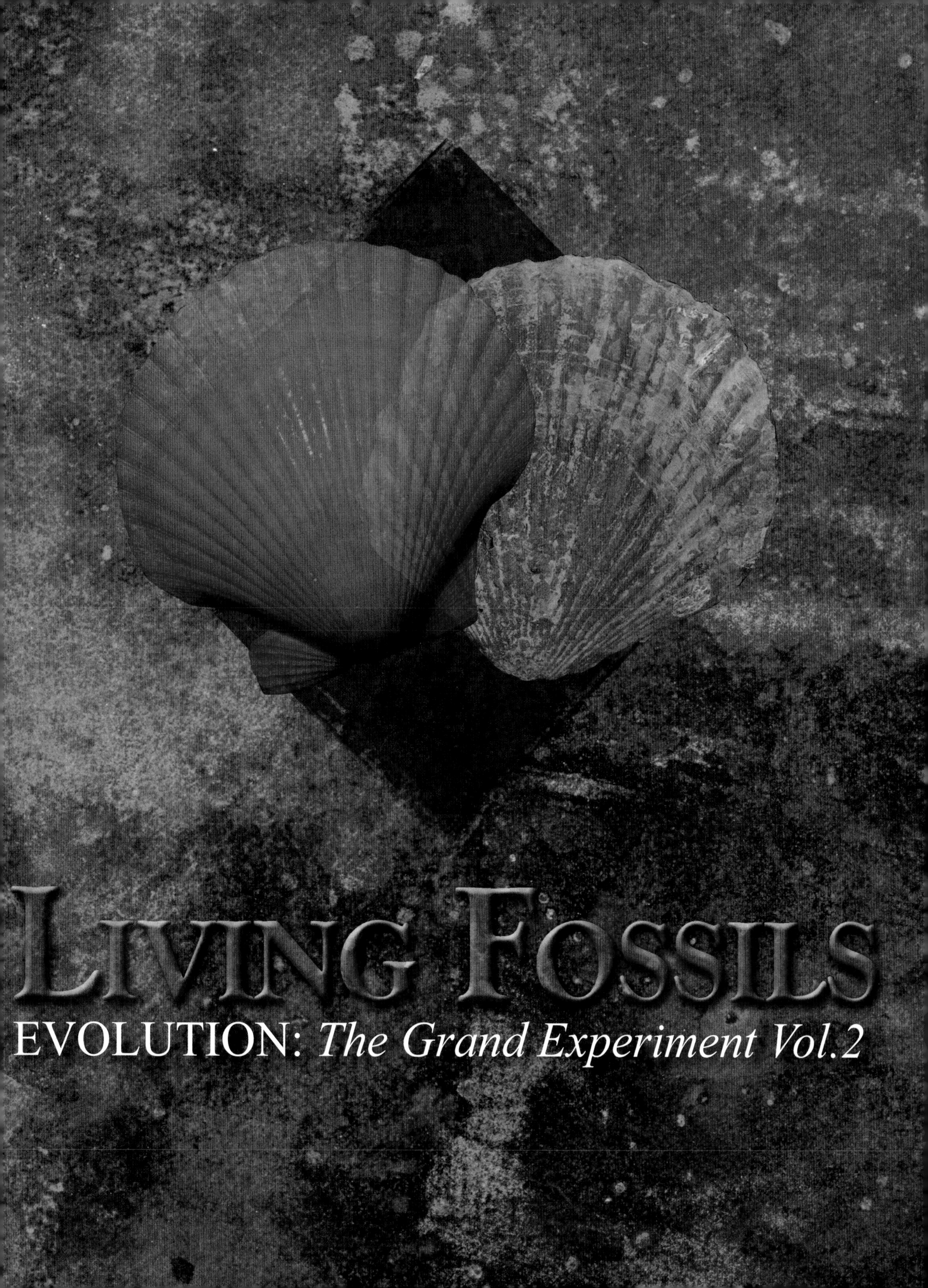
LIVING FOSSILS
EVOLUTION: The Grand Experiment Vol.2

© Copyright 2008, Audio Visual Consultants, Inc.

Graphic Design and Production:
Adriana Naylor, Naylor Design Company
aahtst52@yahoo.com
www.creativeshake.com/adrianaswork

Copy Editors: Carla Azzara, Alon Prunty, and Laura Welch
First printing: February, 2009

Copyright © 2008 by Carl Werner. All rights reserved. No part of this book may be used or reproduced in any manner whatsoever without written permission of the publisher except in the case of brief quotations in articles and reviews.
For information write: New Leaf Press, P.O. Box 726, Green Forest, AR 72638.

ISBN-13: 978-0-89221-691-8
ISBN-10: 0-89221-691-3
Library of Congress Catalog Number: 2008940683

Printed in China

Please visit our websites for other great titles:
www.newleafpress.net
www.a-v-consultants.com
www.TheGrandExperiment.com

New Leaf Press
A Division of New Leaf Publishing Group

Dedicated to those who had the courage to question…

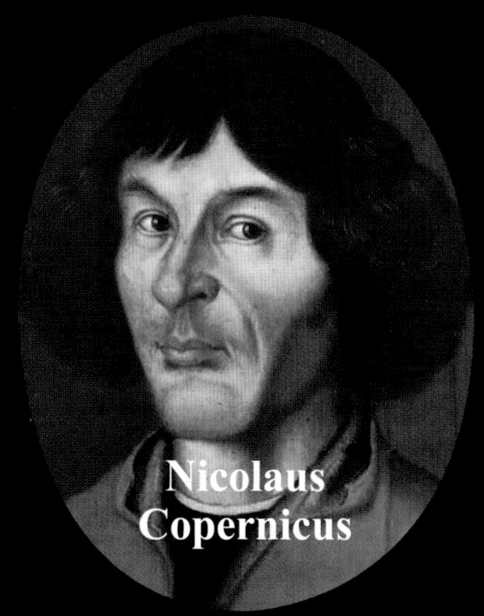

Nicolaus Copernicus

Galileo Galilei

Louis Pasteur

Francesco Redi

… and to those who continue to do likewise.

Dr. Carl Werner

Debbie Werner

About the Author

Dr. Carl Werner received his undergraduate degree in biology, with distinction, at the University of Missouri, graduating summa cum laude. He received his doctoral degree in medicine at the age of 23. He was the recipient of the Norman D. Jones Science Award. He is the author of ***Evolution: The Grand Experiment-The Quest for an Answer*** *and is the executive producer of* ***Evolution: The Grand Experiment*** *video series.*

About the Photographer

Debbie Werner, principal photographer for the book, received her bachelor of science degree from Excelsior College in Albany, New York. She is the principal videographer and producer of ***Evolution: The Grand Experiment*** *video series. She is an avid naturalist and the wife of Dr. Carl Werner.*

Many perceive the fossil record as the greatest proof for the theory of evolution — that life in the far distant past was comprised of strange and unusual animals and plants that evolved over time into the familiar life-forms that surround us today.

Many who oppose evolution argue that plants and animals have not fundamentally changed over time. Although some animals, like dinosaurs and flying reptiles, have gone extinct, the others have remained substantially the same since their origin. These viewpoints radically contradict one another.

In order to solve this controversy, I have reviewed thousands of fossils from the dinosaur era — fossils from Triassic, Jurassic, and Cretaceous rock layers — and compared them to the animals and plants of today. According to one reviewer, you will not find a more complete picture of life during the dinosaur era anywhere.

At the advice and encouragement of others, this book will take a slightly different approach from Volume I. In this volume, I have chosen to tell my personal story, which began with a bet and resulted in a 30-year journey played out over three continents.

Perhaps many of you have already read *Volume I, Evolution: The Grand Experiment-The Quest for an Answer*, an overview of evolution using an easy-to-read textbook format. If so, I certainly hope you learned some new and interesting facts. If not, I encourage you to do so since each volume is a stepping stone to the next.

To help you develop a deeper understanding of this topic, an ongoing video series, *Evolution: The Grand Experiment*, is also available, as well as a teacher's manual and presentation CD for each volume in the series.

Carl Werner

Naming the Book and Video Series

After years of travel and research, one of my first endeavors was to write Volume I of this book series and produce a set of videos on evolution. After months of editing the video series, I nervously took my "final" cuts to local television producers Al Frank (a former local CBS news affiliate), Dianne Becker (producer of *Journey into the Amazon*), and Peter Drochelman (president of the Drochelman Group) for their review. They courteously watched the first production and then shockingly said half of the story was missing! In their opinion, not only did our discoveries need to be told, but our personal story as well. They suggested that any audience would want to know: Who are Carl and Debbie Werner? Why are they spending their own time and money traveling around the world investigating the theory of evolution?

At first, Debbie and I were appalled at their suggestion because we wanted to remain anonymous — so much so that we planned to use pseudonyms. At their urging, we then met with Los Angeles television writer and producer Brian Bird (*Touched by an Angel* and *Call Me Klaus*) and his friend and television writer Chris Easterly. They kindly wrote a pilot for our video series but this time inserted our personal story. When it came time to name the series, Chris Easterly chose the title *Evolution: The Grand Experiment*. When I asked him why he chose this title, he told me he was impressed with the fact that two people were carrying out a "grand experiment," researching and testing the theory of evolution on their own. He convinced us that our journey was as intriguing as our discoveries.

Acknowledgements

Thanks to the following institutions for allowing us access to their collections especially: Harvard Museum of Paleontology, Boston, Massachusetts; Carnegie Museum of Natural History, Pittsburgh, Pennsylvania; Royal Belgian Institute of Natural Sciences, Brussels, Belgium; The Redpath Museum, Montreal, Quebec; Warwickshire Museum, England; Dinosaur Provincial Park, Alberta, Canada; California Academy of Sciences, San Francisco; The Down House, England; Houston Museum of Natural Science, Texas; Chicago Field Museum, Illinois; Museum Victoria, Melbourne, Australia; Palaeontological Museum, University of Oslo, Norway; Dinosaur National Monument, Vernal, Utah; University of California Museum of Paleontology, Berkeley; South Australian Museum, Adelaide; Museum of Geology - South Dakota School of Mines and Technology, Rapid City; New Mexico Museum of Natural History and Science, Albuquerque; Wyoming Dinosaur Center, Thermopolis; Geological Museum, University of Wisconsin, Madison; University of Nebraska State Museum, Lincoln; Milwaukee Public Museum, Wisconsin; Jura Museum, Eichstatt, Germany; Sam Noble Oklahoma Museum of Science and Natural History, Norman; University of Missouri School of Veterinary Medicine; Mesalands Community College's Dinosaur Museum, Tucumcari, New Mexico; Paleontology Museum, Ghost Ranch Conference Center, Abiquiu, New Mexico; Petrified Forest National Park, Arizona; Natural History Museum, Humboldt State University, Arcata, California; Royal Tyrell Museum Field Station, Alberta, Canada; Mississippi Museum of Natural Science, Jackson; University of Wyoming Geological Museum, Laramie; Iguanodon Museum, Bernissart, Belgium; Living World Dinosaur Studios, St. Louis, Missouri; and The Solnhofen Tile Quarry, Solnhofen, Germany.

We would like to thank the zoological parks, botanical gardens, and aquariums for so graciously allowing us to film and photograph at their institutions. The work of those who maintain living examples of plants and animals for all to see is invaluable. Thanks especially to the Sydney Aquarium, Australia; Parndana Wildlife Park, Kangaroo Island, Australia; Muir Woods National Monument, Mill Valley, California; Alaska Sea Life Center, Seward, Alaska; Rainforestation, Cairns, Australia; Missouri Botanical Gardens, St. Louis; Milwaukee County Zoo, Wisconsin; St. Louis Zoo, Missouri; Reptile Gardens, Branson, Missouri; and the World Aquarium, St. Louis, Missouri.

We would also like to thank the museum artists, paleontologists, curators, and fossil preparators who worked to collect, prepare, and display the fossils that we photographed for use in this book. The process of displaying a single museum-quality specimen takes up to ten years. Their efforts and those of others involved in this process, including the museum administrative staffs, are greatly appreciated.

We would also like to thank the scientists who were interviewed for this book especially: Dr. Zhe-Xi Luo, Carnegie Museum of Natural History, Pittsburgh, Pennsylvania; Dr. John Long, Museum Victoria in Melbourne, Australia; Dr. Peter Crane, Royal Botanic Gardens in London; Dr. David Weishampel, Johns Hopkins University; Dr. William Clemens, University of California-Berkeley; Dr. Brint Breithaupt, University of Wyoming Geological Museum, Laramie; Dr. Thomas Rich, Museum Victoria, Melbourne, Australia; Dr. Donald Burge, College of Eastern Utah Prehistoric Museum, Price, Utah; Dr. Thomas Williamson, New Mexico Museum of Natural History and Science, Albuquerque; Dr. Gary Morgan, New Mexico Museum of Natural History and Science; Dr. James Kirkland, Mygatt Moore Quarry, Grand Junction, Colorado; Dr. Paul Sereno, University of Chicago; Dr. Mary Dawson, Carnegie Museum of Natural History, Pittsburgh, Pennsylvania; and Dr. Daniel Gasman, City University of New York (CUNY).

Thanks to those opponents of evolution who were interviewed for this book, either formally or informally, including: Ken Ham, Dr. David Menton, Dr. Georgia Purdom and Dr. Terry Mortenson of the Creation Museum, Kentucky, Dr. Andrew Snelling, and Dr. Duane Gish, Institute for Creation Research, California.

Special thanks to those who helped us identify the species names of organisms, including: Dr. Gordon Hendler, Curator of Echinoderms, Natural History Museum of Los Angeles County; Dr. Peter Sheehan, Head of Geology Department at the Milwaukee Public Museum and Adjunct Professor, Department of Geosciences at the University of Wisconsin, Milwaukee; Dr. Robert Stone, National Oceanic Atmospheric Administration - National Fisheries Service, Auke Bay Laboratories, Juneau, Alaska; Edward Spevak, Curator of Invertebrates, St. Louis Zoo, Missouri; Leonard Sonnenschein, President, World Aquarium and Conservation for the Oceans Foundation, Missouri; Bill Parker, Vertebrate Paleontologist, Division of Resource Management, Petrified Forest National Park, Arizona; Fred Hammer, Education Coordinator, Field Station Visitor Center, Dinosaur Provincial Park, Alberta, Canada; Mr. Bobby Colley, Teaching Support Specialist, University of Missouri School of Veterinary Medicine, Columbia; Paul Cook, Louisiana Department of Wildlife and Fisheries; Martin Bourgeois, Shrimp Program Manager for Louisiana Department of Wildlife and Fisheries; Jody David, Crawfish Project Manager for Louisiana Department of Wildlife and Fisheries; Michael Seymour, biologist for Louisiana Department of Wildlife and Fisheries; and Mr. Jon Wiebe, Louisiana Department of Wildlife and Fisheries.

Thanks to those who contributed photos for this book, especially: Dr. Svetlana Belorustseva, Moscow State University, Russia; Julian Robinson, Australia; Pierre-yves Landover, France; Thomas Wenneck, Norway; Teruo Okamoto, Saitama, Japan; Johnny Jensen, Denmark; Joel Reynaud, Laboratoire de Botanique, Lyon, France; Bill Barss, Oregon Department of Fish and Wildlife; Robert Perry, Malibu, California; David Hansen, University of Minnesota Agricultural Experiment Station; Peter Batson, Fiordland, New Zealand; Jeffrey Humphries, Clemson University, South Carolina; Dr. Jonathan Radley, Warwickshire Museum, England; Laurie Campbell, Scotland, UK; Dr. Richard A. Paselk, Humboldt State University; Keoki Stender, Hawaii; Tosh Odano of Valley Anatomical Preparations; Dr. Paddy Ryan, Thornton, Colorado; Mike Clayton, University of Wisconsin, Madison; Bill Lea, United States Department of Agriculture Forest Service, Southern Research Station; John Seiler, Department of Forestry,

Virginia Tech; David Magney, Environmental Consulting; Dennis Werner, JC Raulston Arboretum, North Carolina State University; Walter Warriner, Consulting Arborist, Redondo Beach, California; Janet and John Garrett, Ipswich, UK; George Sly, Union High School, Dugger, Indiana; Ian Skipworth; James Ownby; Andy Pearson; Nubar Alexanian; Rod Rolle; Ed Bowlby; and Anne Muecke. (Please see Photo Credits.)

Debbie and I would like to thank the governmental institutions overseeing the dig sites, museums, and national parks where we filmed, including The Bureau of Land Management, Grand Junction, Colorado; the governments of Australia, Belgium, Canada, Germany, Mexico, United Kingdom, U.S. Virgin Islands, and the United States.

We feel a great indebtedness to Carl's writing assistant and copy editor, Carla Azzara, whose writing and editorial skills helped to transform his ideas into written form. Her array of abilities have been invaluable in guiding this project to completion. For this, we will be forever grateful.

Also, we would like to acknowledge our graphic designer, Adriana Naylor, for her fabulous work. She continues to amaze us with her fresh designs and layouts for this book series. By taking our photographs and stories and placing them on rich backgrounds, she makes the entire book light up.

To our friend, Alon Prunty, whose adept editorial skills have been instrumental in preparing this book for presentation, adding the final touches. Our sincere thanks to you.

The staff members at New Leaf Publishing are the unsung heroes in this book series. They have transformed our books into marketable products and have been able to garner sales from all over the world. I am indebted especially to Tim Dudley, President of New Leaf Publishing; Lydia King, Chief Financial Officer; Laura Welch, Editor-in-chief; Stacey Drake, Publicist; Diana Bogardus, Head of the Graphic Department; Graphic Designers Terry White, Rebekah Krall, Janell Robertson, Judy Lewis, Bryan Miller, Brent Spurlock, and Jeff Patty; Aubrey Peden, Accounting; Janell Robertson, Marketing; Dee Oehlerking, Customer Service; Amanda Price, Assistant Editor; Don Enz, National Sales; Josh Shields and James Davis, Sales; and last, but certainly not least, the warehouse crew—Debbie Ratliff, Wes Ratliff, Micaela Sciacca, and Tonie Jackson.

Lastly, thanks to our families, friends, and board of advisors who have helped us pursue this dream, especially the Werners, the Lawsons, the Huberts, the McClures, the Wards, the Tichaceks, the Williams, the Ottos, the Bachmanns, the Carrons, the Puhses, the Deters, and the Welshs. Thanks to the following who reviewed the manuscript: Aaron Artz, Kelly Mooney, George Tichacek, Fran Werner, and David and Heidi Frick.

CHAPTER 1

CHAPTER 2

CHAPTER 3

Chapter 9

Chapter 10

Chapter 11

Chapter 12

Chapter 22

Chapter 23

Chapter 24

CHAPTER 25

Appendices

Glossary

Bibliography

Photo Credits

Index

Teacher's Manual

Volume III

The Challenge That Would Change My Life

My lifelong interest in evolution began with an innocuous challenge over dinner.

Chapter 1

My Story Begins

It is said to have a proper story, three essential components are required: a beginning, a middle, and an end. While I will adequately provide you with a beginning (my story begins with a bet in medical school), and a middle (which leads to an incredible 30-year journey), I hesitate to say that my story has an "ending" for two reasons. First, I have more information I would eventually like to share with you (Volumes III and IV of this series); and second, my version of an ending doesn't really matter. As the author, I would prefer you write the ending. When you finish, you need to ask yourself: Has my perception of the past been changed?

Before I get into the specifics of *how* I became fascinated with the theory of evolution, you need to understand my background.

I was born in 1959 in a large Midwestern city and raised Catholic. I attended Catholic grade school and high school. Through my early years, I believed in the creation story, and the Bible stories of Adam and Eve, Noah, Moses, etc. I had no reason to doubt them and of course, no one gave me any other options.

It was in my later high school years, between the ages of 15 and 17, that I found myself drifting away from my religious ideas and beliefs. This was my state of affairs when I was accepted to an accelerated college and medical school at the relatively young age of 17.

My first class in med school was physiology. Here, the professor taught us the evolutionary principle of "Ontogeny Recapitulates Phylogeny" created by Dr. Ernst Haeckel in the late 1800s.

I had never heard of this concept and neither could I pronounce it. Fortunately, the professor had the class repeat the phrase "On-todge-en-knee Re-ca-pit-you-lates Fi-lodge-in-knee" over and over until we could say it smoothly and efficiently like a machine gun spitting out bullets

Above: *Author as a teenager on hospital rounds in medical school. (Author is second from left.)*

at a thousand rounds per minute. He proceeded to explain what it meant: Prior to birth, animals retrace the history of evolution in their embryonic stages. For example, humans had their origin in a single-cell bacterium, which evolved into an invertebrate like a jellyfish, then a fish, then an amphibian, a reptile, a mammal, a monkey with a tail, and finally a tailless ape. He then showed us Dr. Haeckel's drawings of human embryos in various phases of development, such as a single-cell fertilized egg (similar to a single-cell bacterium), an embryo with "gill slits" (similar to a fish with gills), and an embryo with a tail (similar to a monkey). On the next page are two of Haeckel's drawings.

These drawings were extremely compelling to me, especially the "fact" that humans had gills and a tail. After this lecture, I found myself rapidly accepting evolution.

Years later, I learned that the drawings used to demonstrate Ontogeny were extremely inaccurate. "When critics brought charges of extensive retouching and outrageous fudging in his famous embryo illustrations, Haeckel replied he was only trying to make them more accurate than the faulty specimens on which they were based." [1]

Here are some of Haeckel's errors: (1) Dr. Haeckel made the images of different animal embryos look similar even though the embryos do not appear this way in life; (2) Haeckel referred to neck pouches in the human embryo as "gill-arches," yet there are no fish gills in the human embryo; and (3) Dr. Haeckel referred to the end of the vertebral column of the human embryo as "a tail" even though these vertebrae coincide with the sacrum and coccyx to which the pelvic organs are attached.

Sadly, I cast my vote for evolution in 1977 based on this faulty evidence. [2] No one in my medical school told me that Haeckel's drawings were shown to be inaccurate 80 years earlier. [3,4] Even sadder is the fact that Haeckel's drawings are still part of some medical school textbooks today.[5] (See interview below.)

Dr. Ernst Haeckel
1834-1919

Left: *Dr. Ernst Haeckel falsified embryo drawings to support his evolutionary theory of Ontogeny Recapitulates Phylogeny.*

Right: *Haeckel's drawings were used in my first physiology class in medical school even though they were shown to be erroneous in the late 1890s.*

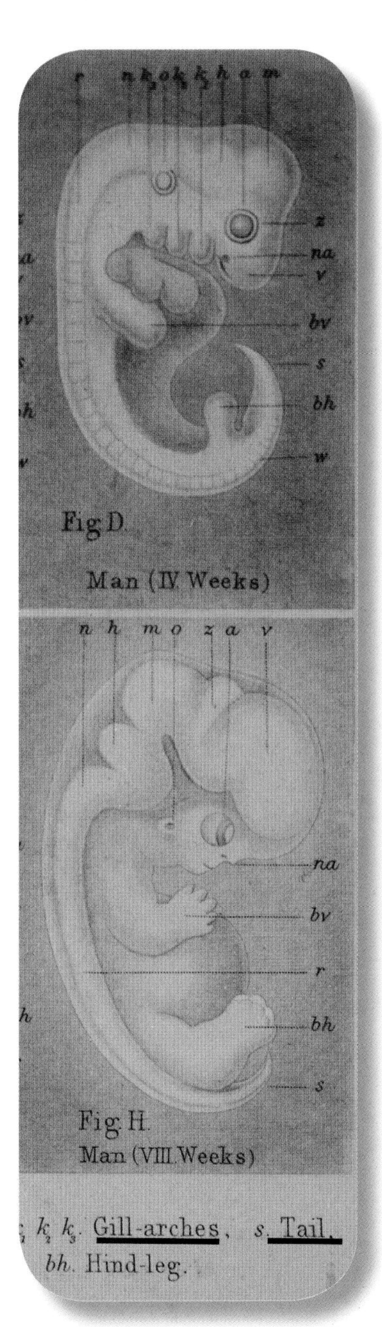

Interview from *Evolution: The Grand Experiment* video series

*"The accusation that Haeckel had fraudulently portrayed embryos in the latter part of the 19th century was an accusation that was raised at the time....**Many of the medical textbooks today still duplicate the erroneous drawings** that Haeckel had portrayed in the 19th century."* [5]

Dr. Daniel Gasman, professor of history, City University of New York (CUNY). Dr. Gasman is considered an expert on Haeckel.

"But there's always been a couple of problems with the big bang theory. First, when you squeeze the entire universe into an infinitesimally small, but stupendously dense package, at a certain point, our laws of physics simply break down. They just don't make sense anymore." [6]

— Dr. Greene

Dr. Brian Greene is professor of mathematics and physics at Columbia University.

Four Questions

One year later, in the middle of my sophomore year of college, I went out for pizza with a classmate. In my mind, it was just a social time to chew the fat. While eating dinner, we talked about our classes and friends. Then, for some unknown reason, my classmate began to ask some serious and pointed questions — questions that would forever change my life.

Q: *What did I think about evolution?*
A: *I believe.*

Q: *What did I think about the problems with the fossil record which cast doubt on the theory of evolution?*
A: *I didn't know there were "problems" with the fossil record.*

Q: *What did I think about the problems with the laws of physics in the big bang model?*
A: *I don't know. I had never heard of "problems" with the laws of physics in the big bang theory.*

My friend's last question sunk me. It pertained to an area I was very familiar with, biochemistry.

Q: *How could life begin if proteins do not form naturally?*

I thought to myself: *"He's got me."* I had studied the chemical equations of proteins and aced them in class, but I had never applied them to the origin of life. Let me explain. The theory of evolution suggests that the very first form of life, a single-cell organism, formed spontaneously (or naturally) out of chemicals. But proteins, one of the necessary components for a single-cell organism, do not form naturally. How could life begin if proteins do not form naturally out of chemicals?

A seed of doubt entered my mind that day, and I felt a wave of emotion as I wondered, *"Have I been duped into believing evolution?"*

"*The* [physics] *formulas we use* [in the big bang theory] *start giving answers that are nonsensical. We find total disaster. Everything breaks down, and we're stuck.*" [6]

— Dr. Gross

Dr. David Gross was the recipient of the Nobel Prize in Physics in 2004. He is the chair of theoretical physics at the University of California, Santa Barbara.

"No one has ever seen or witnessed a protein molecule form naturally." [7]

— Dr. Gish

Dr. Duane Gish opposes evolution. He received his Ph.D. in biochemistry from the University of California, Berkeley.

The Challenge

Before I could gather an adequate response to the protein problem, my friend fired his last salvo. *"Carl, I challenge you to prove evolution."* I retorted *"That's crazy. It has been proven!"* But he had made his mark. His verbal shot lodged in my brain like a bullet. I thought to myself, *"How could evolution be true if one cannot reconcile these important issues?"*

His points concerning the formation of proteins and the laws of physics seemed believable, but I wasn't quite sure I trusted my classmate's lofty accusations that there were "problems" with the fossil record. How did he know? This was my med school buddy talking, not a paleontologist. He told me that nearly all the animal groups have missing links in their evolutionary history, despite finding millions and millions of fossils. How could this be? I had always assumed the so-called missing links (the fossils portraying one animal type changing into another, such as a dinosaur changing into a bird) are missing because the fossil record was poor. He pointed out the other logical possibility — that the proposed missing links never existed and that was why they had not been found. His reasoning seemed plausible. Still, I was not convinced. I am, by nature, skeptical. But because of the simplicity and eloquence of his arguments, I gave them some credence.

Now I was unnerved. How could there be such fundamental problems with the big bang theory, the origin of life, and the fossil record if evolution was true?

With this casual challenge began the adventure of a lifetime, to prove evolution right or wrong. I decided I would review the evidence for the theory of evolution from top to bottom and then devise ways to test it. I felt up to the task because I had been afforded valuable experiences in science and

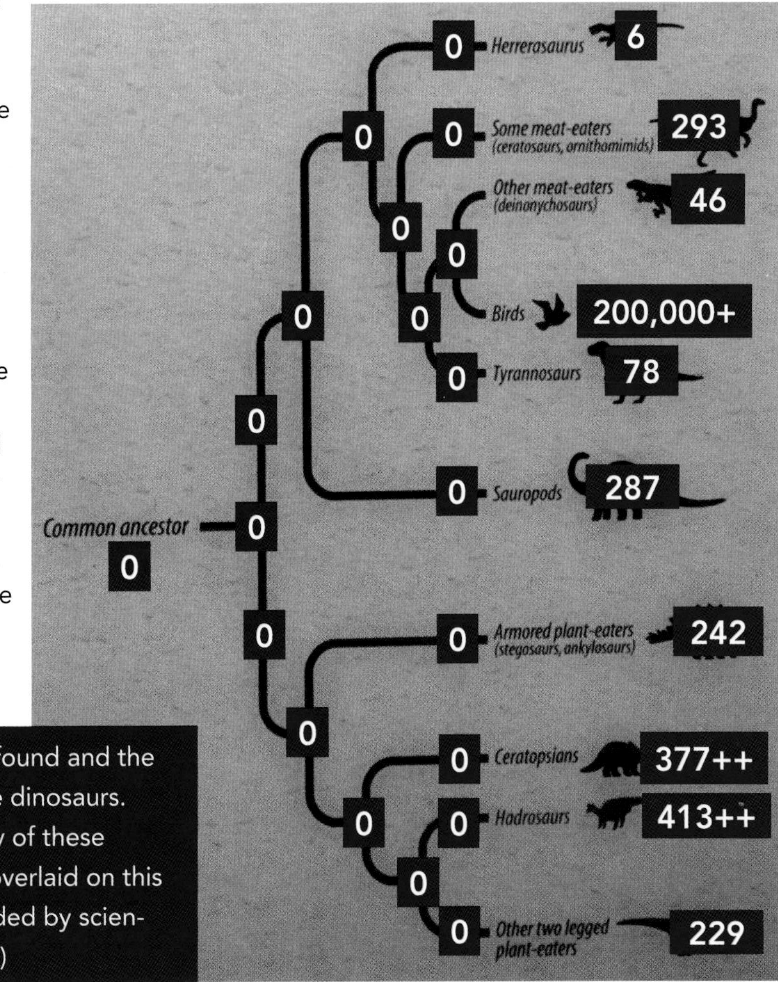

This chart shows the number of dinosaurs found and the number of evolutionary ancestors for these dinosaurs. There is not a single direct ancestor for any of these dinosaur species or groups. (Information overlaid on this Chicago Field Museum diagram was provided by scientists who support evolution. See Volume I.)

experimentation. From all of these experiences, I learned how to apply the scientific method used to prove or disprove an idea.

By the time I accepted the challenge in my sophomore year of college, I had been educated in chemistry, organic chemistry, biochemistry, genetics, anatomy, physiology, embryology, and biology. My intention was to study evolution in my free time and hopefully wrap this up in a few years. Little did I know it would take decades, studying and traveling, to arrive at a definitive answer.

Above: *Dr. Werner is the medical director of an emergency room near St. Louis. Over the years, he has held the positions of director and assistant director of the emergency departments at several other hospitals.*

ST. LOUIS POST-DISPATCH

FINAL

Carter Promises Fight For Yearly Fuel Tax

Carter Seeking Aid For Energy Plan

Student Uses Drug

friday

Above: *Author featured on the front page of the St. Louis Post-Dispatch in 1977 after winning the Norman D. Jones Science Award for his experiment dealing with the food preservative EDTA. For his work, he was awarded a four-year academic scholarship to the University of Missouri where he obtained a degree in biology, with distinction, graduating summa cum laude. Later, he received his doctoral degree in medicine at the age of 23 and scored 99th percentile (out of 99) on his clinical medical boards.*

Now most people would find it difficult to believe that someone would go on a lifelong quest stemming from an innocuous challenge over dinner. Yet, this is all rather telling about me. I am an independent thinker and a seeker of truth. Over the last 30 years, I have to confess, there were times I wished that conversation had never happened. I would have led a "normal" life as an ER physician, with more time to enjoy my favorite sports of fishing and sailing. But the reality is you cannot go back and change the past.

How Can You Verify Evolution?

This chapter describes how the author formulated his first experiment to test evolution.

Chapter 2

Evidences for and against Evolution

Given the enormous complexity of the theory of evolution, I began to contemplate how to dissect and evaluate the theory. I started with what I knew at that moment. There were many evidences for evolution *but* there were also many evidences against it. Here were my lists:

Charles Darwin, 1809-1882, father of the theory of evolution.

Evidences *for* Evolution

1. ***Ontogeny Recapitulates Phylogeny.*** When I took on the challenge, this was the best evidence for evolution that I had ever heard.

2. ***Most scientists believe it.*** Most university professors of biology and paleontology would offer their support for the theory of evolution without reservation. However, just because scientists believe a theory does not make it true. For example, the majority of scientists once believed the earth was the center of our planetary system, but this is false. The sun is the center. The majority also thought spontaneous generation was the explanation for how life began. This theory suggested that mice came from dirty underwear and that maggots came from rotting meat. I could offer other examples, but the point is that science is not a democracy where the majority vote wins. Time tests the validity of scientific ideas.

3. ***Ape-men.*** Given the overwhelming evidence that ape-men existed in the past, how could evolution not be true? Ape-men are the Holy Grail of evolution, the missing links, proving that man evolved from apes. If evolution was not true, then all of the proposed ape-men would be either just apes or just men, including Lucy, *Africanus*, Neanderthal, etc. At the time I accepted the challenge, that was hard to believe. Since then, I have conducted interviews with the anthropologists involved in the recovery of these fossils. (This new information will be addressed in Volume III of this series.)

4. ***Some fossil evidence of animals changing from one animal type into another.*** When I took on the challenge, there were these few well-accepted examples of animals changing from one type into another: the evolution of modern horses from four-toed animals, the evolution of birds from dinosaurs, the evolution of amphibians from fish, and the evolution of mammals from reptiles. In each case, scientists reported finding the connecting links between these animal groups, thereby proving evolution.

Evidences *against* Evolution

1. ***Too many missing links.*** Millions of fossils have been collected by museums, yet the evolutionary ancestors have not been found for most animals and plants. (This was presented in Volume I.)

2. ***The big bang theory does not work.*** When scientists apply the laws of physics to the big bang model, the laws of physics simply do not work.[1,2] To me, this was evidence against a natural, spontaneous beginning of the universe.

3. ***Life could not begin spontaneously.*** Proteins are necessary for life to exist, yet they do not form out of chemical elements, spontaneously, in the laboratory. Without proteins, life could not have begun. (This was presented in Volume I.)

At first glance, it appeared there were evidences both for and against evolution, but you can't have it both ways. If evolution is true, everything should line up under the first list. If evolution is not true, everything should line up under the second list. Remember, Copernicus, Galileo, and others had those few nagging astronomical observations concerning the sun, which they could not reconcile with the earth being the center of our planetary system. By tenaciously clinging to the few pieces of information that didn't fit, they eventually arrived at the correct explanation.

As a college student, I remember looking at these two lists with daunting emotion, knowing that to verify any *one* of these evidences might take years. To test *all* of them might take decades. Already, my plan of having this all wrapped up in a few years was beginning to unravel.

The Scientific Method

In order to test a theory, one must apply the scientific method, defined as the body of techniques used to investigate phenomena, acquire new knowledge, and verify theories.[3]

Using the scientific method, a scientist will start with an idea (theory or hypothesis) and then test the validity of his idea by vigorously trying to *disprove* it. If he or she can't falsify it, then the original theory remains tentatively true.

In my case, if I could not refute the theory of evolution with an extremely tough evaluation, then evolution would remain tentatively true.

For me the stakes were high. If evolution was verified, I could, once and for all, abandon the last vestiges of my religious faith without guilt or reservation. On the other hand, if the theory of evolution was not supported by the evidence, then I had lots of soul-searching to do regarding my beliefs. For me, a left-brained person, it was either black or white. I needed to know the answer — mechanistic evolution or metaphysical creation?

Postscript

Many scientists believe that evolution is true and no experiment could be designed to disprove it, but I disagree. How can anyone be so sure of any theory without evaluating and testing it? I at least intended to apply an honest effort and test the idea — evolution is either supported or not supported by the evidence.

Designing My Experiment

For the next 18 years following my college classmate's challenge, I immersed myself in reading and studying the topic of evolution. I digested books in every pertinent area — geology, biology, and biochemistry — anything that would help me test evolution. At one point, I even opted for a downgrade from the position of medical director of a metropolitan emergency room to a less busy emergency department to allow me more time to read. Yet, there came a time when merely reading about evolution became woefully inadequate. Secondhand information from articles and books was no longer enough. I needed to be in the field, touching the fossils and talking with the specialists. I was now ready to take my investigation to the next level and design an experiment. But which area of evolution would I choose to test first? I thought about this for nearly two decades. Finally, while driving to work one day in 1997, the design for my first experiment clearly solidified in my mind. To say the least, I was excited to come up with a simple test. Let me explain.

Evolution teaches that 1) animals and plants changed dramatically over time, from one type into a completely different type, through random mutations, and 2) in this ever-changing line of animals and plants over millions of years, due to the principle of the survival of the fittest, the weaker predecessors became extinct. I would take the opposite stance in order to test evolution. ***I predicted that if evolution was not true, then animals and plants would not change significantly over time. Accordingly, I predicted I should find fossils of modern animal and plant species in the "older" fossil layers.***

To carry out my experiment on a smaller, more manageable scale, I decided to focus on just one section of the rock layers — the fossil layers associated with dinosaurs, the Triassic, Jurassic, and Cretaceous rock layers. At the time, I did not know of a *single* modern plant or animal living during the dinosaur era, except for a few well-recognized so-called "**living fossils.**" **Living fossils are fossils which look very similar to modern plants or animals.**[4] But even these dinosaur-era "living fossils" — the dragonfly, the garfish, the coelocanth fish, and the horseshoe crab — were so different from the living forms they were assigned different

Above: *The Grand Canyon*

genus and species names indicating they could not reproduce with modern forms (if they lived at the same time).

Because I did not know of any modern species of animals or plants found in dinosaur rock layers, my experiment seemed destined to fail from the start. But I was determined to plow ahead.

A Few "Slight" Details Remaining

Now there were a few "slight" details remaining: How would I pull this whole thing off? In order to test my idea, I would have to gain access to and photograph the fossils at the various dinosaur dig sites and museums throughout the world and compare them to modern forms. How would I do this? Even more important, how would I enlist my wife's support? I thought about this for a few months, and finally, after taking a very deep breath, I made a proposal to my wife. I asked her if she was willing to make "a trip out West" in order to further my research of evolution and possibly write a book.

I was afraid that all of my research up to this point would be for naught and it would all come down to one terse, "Forget it." Frankly, her answer surprised me.

Now, Debbie knew I was becoming exceedingly frustrated in concluding my thoughts about life by merely reading books and articles, so she agreed, but with one stipulation. She explained to me that she was not sure anyone would believe my results, without some sort of *proof* for my findings, no matter what they were. *"Why would anyone take your word?"* she asked. *"What good would it do for you to go out to these dig sites and see these fossils? Who would believe you? Even if you did uncover something, how could you prove your findings one way or another to a doubting audience?"* She suggested that if I were going to do this, I better do it right and document everything so **we** weren't wasting our time. When she said *"we,"* I knew she had thrown herself in the ring and was willing to go with me.

I now found myself in a unique position to carry out this project, this quest. Everything was falling into place. I had a wife with many talents. I wasn't rich, but I had enough money to independently fund the project. In order to gain access to the museums and scientists and to document my findings, I formed a natural history television and book production company (Audio Visual Consultants, Inc.), charged with finding answers to the deepest scientific questions concerning the origin of life. Soon we were buying high quality photographic equipment, along with television quality video equipment. Before our first trip, we began practicing, photographing dime-sized objects and filming television-quality interviews. In six month's time, we were ready to start out on an itinerary that covered dinosaur dig sites from Denver, Colorado, to Alberta, Canada.

Little did I realize that this "trip out West" was just the beginning. Eventually, we trekked 108,561 miles over three continents — three times the number of miles that Charles Darwin traveled on his famous HMS Beagle trip. I had no idea we would invest such an enormous amount of time and money in all this.

I know this may sound somewhat unbelievable. But as I described myself earlier, I am very skeptical and curious. I had to find the answer.

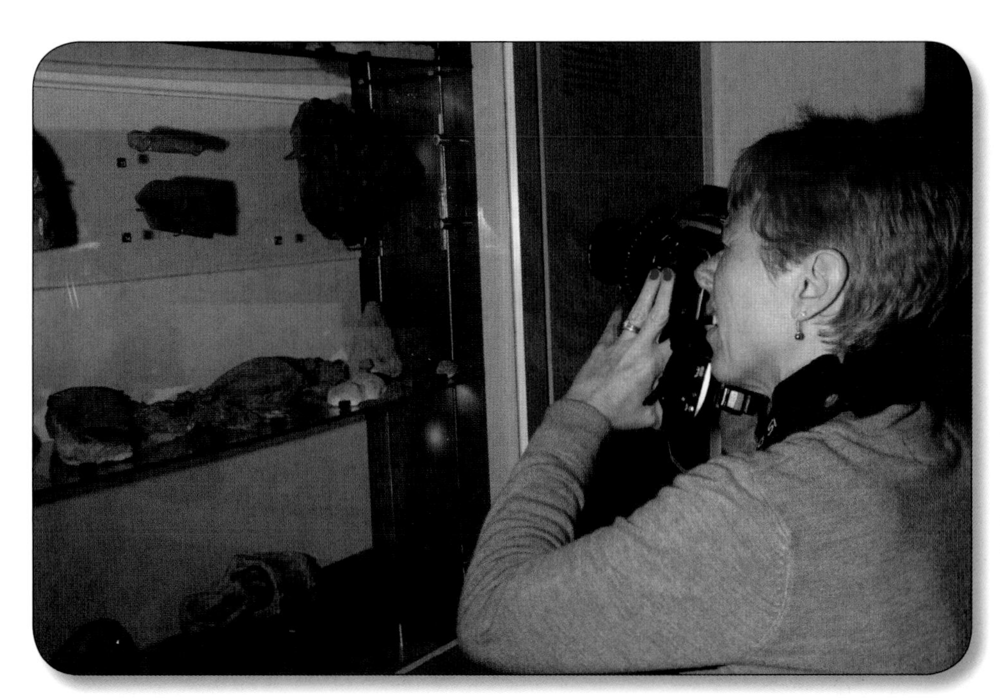

Left: *Over time, our equipment and techniques became more sophisticated. Here Debbie is photographing fossils with a detached flash at the Museum Victoria in Melbourne, Australia. (Her flash is lying on her shoulder as she checks focus and frame.)*

Right: *By the time we reached our 80,000th mile, we began to grow tired and hired a second film crew to help us. Here we are filming an interview at the Carnegie Museum with Debbie running one camera and a local film crew running a second camera and sound.*

"I had no idea it would lead to all of this. We trekked 108,561 miles over three continents to finish the challenge."

Our Journey Begins

Later that year, in 1997, we headed out West. Our plan was to fly to Colorado, rent an RV, and then drive to the dig sites and museums from there. It was a good plan except for one slight hitch — the astronomical rental cost ($175 to $250 a day plus gas) for a 25-foot RV. That was a lot of money. Upon learning this, I asked Debbie to shop around for something less expensive. After further research, she found a whole line of "economy rental RVs." Over the phone, we settled for one at $125 a day, and yes, the old adage still holds true: We got what we paid for.

At the Denver airport terminal, I knew we were in trouble when we climbed into the RV and it looked like something out of the 70s. We could live with that. But as the young lady from the RV rental company began to show us the "features" of the RV, we realized we were dealing with more than a cosmetic problem. We soon learned the refrigerator wouldn't operate. (This was an issue because my wife is addicted to diet Coca-Cola.) Then the RV company representative tried to show us how to check the engine's oil level, but she couldn't get the hood open. (Later, we couldn't drain the sewage tank.) I won't bore you with the other details, but this RV, which affectionately came to be known as "Betsy," was a bomb.

Because I had booked our first interview for later that day at the University of Wyoming in Laramie, and the RV rental store was 60 miles in the *opposite* direction, we didn't have time to switch rental RVs. We had no choice but to accept the refined piece of tin, sign the rental papers, and move on. On that note, we began our journey. I couldn't help but wonder what other surprises lay ahead.

Several days later, we arrived at the Wyoming Dinosaur Center in Thermopolis, the first dig site on our itinerary. My experience there — touching and examining actual fossils — was incredible. After years of preparation, I was now able to begin my experiment by asking this one simple question: ***"Have you found any fossils of modern animal or plant species at this dinosaur dig site?"*** But getting an answer wasn't as easy as I had hoped.

Above: *One of the dozen scientists we interviewed on our "trip out West," Dr. James Kirkland at the Mygatt Moore Dinosaur Quarry.*

Above: *The famous one-hub-cap "Betsy" in her glory.*

Learning *How* to Ask

Debbie and I both shared in the interviewing process at the various dig sites. She ran the video and audio equipment as I conducted the interviews. At times, she would purposefully play the role of the "dumb blonde" and ask the most difficult questions. Other times, she would chime in with incredibly intuitive questions.

Right from the start, my question (concerning the possibility of modern plant and animal species living with dinosaurs) met with resistance. I soon discovered that the scientists at the dig sites were staunch supporters of the theory of evolution and for them, finding "modern" life in ancient rock layers was in opposition to their beliefs. One scientist even refused to be interviewed after reading my questions. I quickly realized that getting answers for my experiment was going to be difficult, to say the least.

After many frustrating encounters, I finally got a break while interviewing Dr. William Clemens from the University of California at Berkeley. At the time, Dr. Clemens was embroiled in a huge scientific argument over the topic of asteroids killing off the dinosaurs. When I asked Dr. Clemens why he didn't believe the idea that dinosaurs went extinct as a result of an asteroid hitting the earth, his answer astounded me. He said that if an asteroid impacted the earth and killed off the dinosaurs, it should also have killed off the butterflies, bees, frogs, and salamanders that were living with the dinosaurs. *WHAT*? This was the first time I had been told that butterflies, bees, frogs, and salamanders were living during the time of the dinosaurs! He did not say these were modern species, but I wondered if they could be nearly the same as those living today.

From that point forward, in order not to offend anyone and to get the information that I wanted, I radically changed my approach and posed this question to the scientists: *"Some scientists, such as Dr. Bill Clemens, have suggested an asteroid did not cause the extinction of the dinosaurs because environmentally vulnerable animals, like salamanders and butterflies, lived through this dinosaur extinction event. At this site, where you are working, have you found any animals that survived the dinosaur extinction event — any modern-appearing animals that are still alive today?"*

The query, as revised, did not specifically challenge the theory of evolution, rather, it focused on dinosaur extinction. And because I switched my choice of words from "modern" to "modern-appearing," scientists were now much more relaxed, open, and willing to talk. In these conversations, I was given examples of modern-appearing animals and plants that were found at dig sites alongside dinosaurs; yet, they always pointed out that the ancient fossils had different genus and species names.

By framing my question in this manner, my list of "*modern-appearing*" animals and plants quickly grew. Still, I didn't know what this all meant since I predicted finding "*modern*" plant and animal species with dinosaurs, not "modern-appearing."

It wasn't until later that I realized that something was awry with the *naming* of species — a questionable and observable pattern which I call "The Naming Game."

Above: *Dr. William Clemens, University of California, Berkeley. Our interview with him gave me the break that I was looking for.*

Above: *Setting up for an interview at the South Australian Museum in Adelaide.*

The Naming Game

This chapter describes the subjective nature of assigning species names and describes how this could falsely lead one to conclude that animals and plants changed dramatically over time.

Chapter 3

"I began to have an uneasy feeling that species names were being incorrectly assigned to the fossils."

The Biological Classification System

Before I explain what I refer to as "The Naming Game," you first need to understand the system used by scientists to classify animals and plants. This classification system is simply a method by which to group all organisms using a hierarchical structure of categories, such as kingdom, phylum, class, order, family, genus, and species. Each of the seven categories is more specific than the previous. For example, a phylum contains all of the classes below it, and a class contains all of the orders below it. An organism is classified as part of a kingdom, then a phylum, then a class, order, family, genus, and lastly a species. The charts below show the scientific classification for a human and a sassafras tree.

Kingdom	Animal
Phylum	Vertebrates
Class	Mammal
Order	Primate
Family	Apes/Human
Genus	Homo
Species	sapiens

Human

For plants, the word "division" is used instead of "phylum". A sassafras tree would be classified this way.

Kingdom	Plant
Division	Flowering Plants
Class	Two Leaves around Each Seed
Order	Laurels
Family	Laurel Family
Genus	Sassafras
Species	albidum

Sassafras

Homo sapiens
(Human)

Pan troglodytes
(Chimp)

Canis lupus
(Wolf)

Canis familiaris
(Dog)

Naming an Organism

Most people recognize an organism by its common name, not its scientific name. The scientific name for any organism consists of two parts, *written in italics*. The first part denotes the genus name and the second part denotes the species name. Looking at the chart on the previous page, the scientific name for a human is *Homo sapiens*, and the scientific name for a sassafras tree is *Sassafras albidum*.

Unfortunately, the system is very subjective and frequently there are disagreements among scientists regarding how to group or classify animals or plants.

Equus zebra
(Zebra)

Equus caballus
(Horse)

The Classification and Naming of Six Animals

Common Name	Human	Chimp	Wolf	Dog	Horse	Zebra
Kingdom	Animal	Animal	Animal	Animal	Animal	Animal
Phylum	Vertebrate	Vertebrate	Vertebrate	Vertebrate	Vertebrate	Vertebrate
Class	Mammal	Mammal	Mammal	Mammal	Mammal	Mammal
Order	Primates	Primates	Meat-Eating	Meat-Eating	Hoofed Odd Toe	Hoofed Odd Toe
Family	Apes/Human	Apes/Human	Dog	Dog	Horse	Horse
Genus	Homo	Pan	Canis	Canis	Equus	Equus
Species	Sapiens	Troglodytes	Lupus	Familiaris	Caballus	Zebra
Official Name	*Homo sapiens*	*Pan troglodytes*	*Canis lupus*	*Canis familiaris*	*Equus caballus*	*Equus zebra*

What Is a Species?

Next, you need to know the definition of a "species." There are many, but the simplest, most straightforward, and most testable definition is this: ***A species is a group of animals that can produce fertile offspring.***[1] For example, if a Beagle (dog) and a Great Dane (dog) mate, they generally will produce a litter of puppies. When these puppies mature and eventually mate, they too can produce a litter of puppies (dogs). Because of this, the Beagle and the Great Dane can confidently be assigned to the same species, *Canis familiaris*, even though they differ greatly in appearance.

Using this same definition, the opposite is also true. Two animals that are not of the same species cannot produce fertile offspring. A cat cannot mate with a dog and produce offspring. They are two different species. The same is true for a mouse and a rabbit. Now a horse, which has 64 chromosomes, may mate with a donkey, which has 62 chromosomes, and produce a hybrid animal, a mule, which has 63 chromosomes. But a mule is infertile and because of this, the donkey and the horse are not the same species; they are usually incapable of producing *fertile* offspring.

Naming Fossil Species

How can you tell if two animals, one fossilized and one living, are the same species or not? If a paleontologist finds a fossil animal in a rock layer *that looks similar to a modern animal*, how should he or she name it? Should the scientist assign the modern species name to the fossil? Or should the scientist assign a new and unique species name?

Unfortunately, the naming of a species is not carried out by a committee of scientists. Usually, the individual scientist who discovers the fossil assigns the species name. In some cases, the scientist who discovers a fossil whole-heartedly supports evolution and his or her choice of a new species name may reflect this bias. Often, a scientist gains recognition from peers by naming a new species. There is an inherent risk of making a mistake in assigning species names due to the subjective nature of the process. Since no one can test the species name assigned to a fossil using the simple test of reproduction, there is always the potential to erroneously assign a fossil a new and unique species name, even when it may look nearly the same as a modern species.

Variations within Species

If a paleontologist was unfamiliar with the variations within living species, it is possible he or she could misinterpret differences and mistakenly give each fossil a new, erroneous species name, based on minor changes only. Taking it one step further, this name change could give the false impression that evolution, or dramatic change, has occurred over time. The only way this can be sorted out is to look at the fossils and living forms.

One must be careful when assigning a new species name since there are many variations or differences in size, shape, and features in any one species.

West Highlander

Pug

Pointer

There are great variations in dogs, yet they are the same species (*Canis familiaris*).

Pomeranian

Doberman

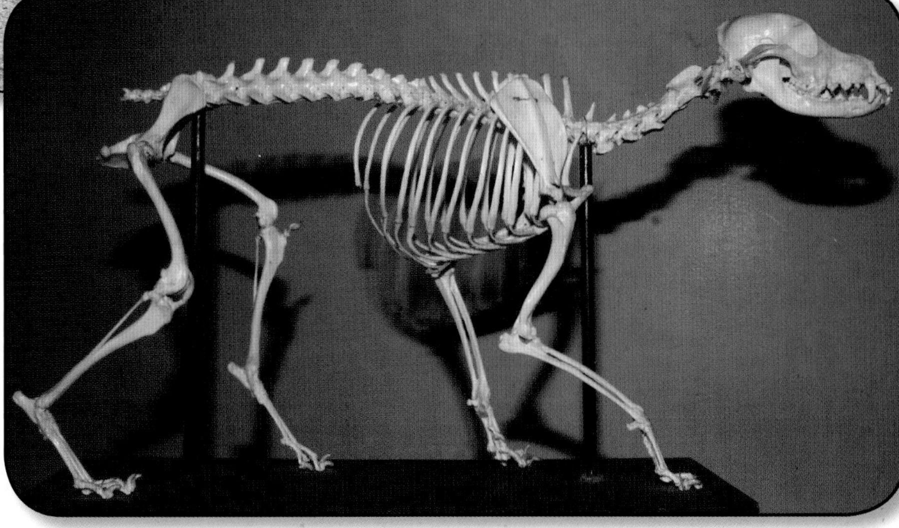

French Mastiff

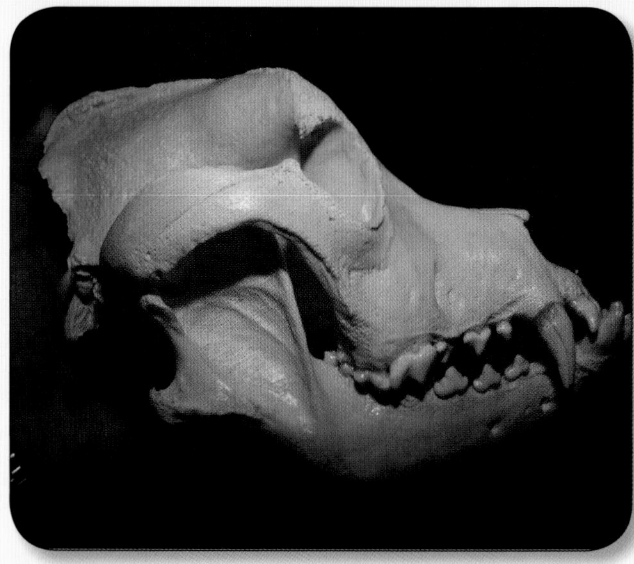

Mastiff Skull

Tremendous Variations in Dog Skulls

Chihuahua Skull

Chihuahua

Irish Wolfhound

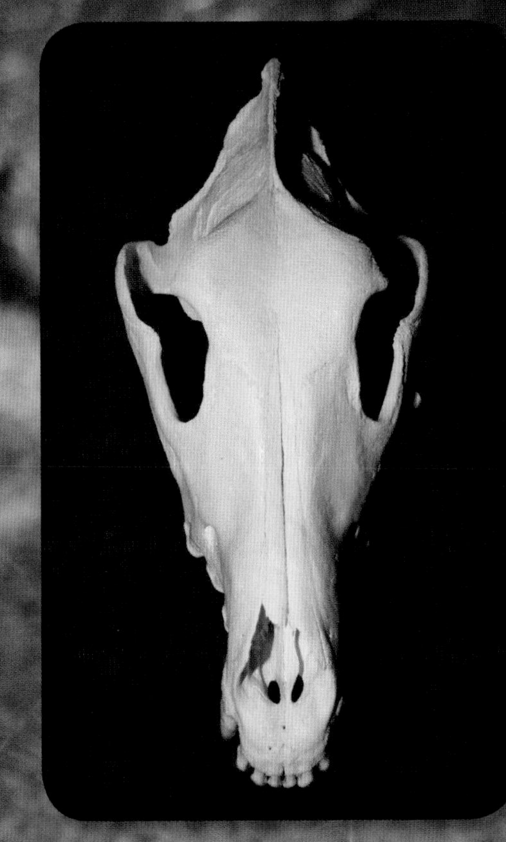

Irish Wolfhound Skull

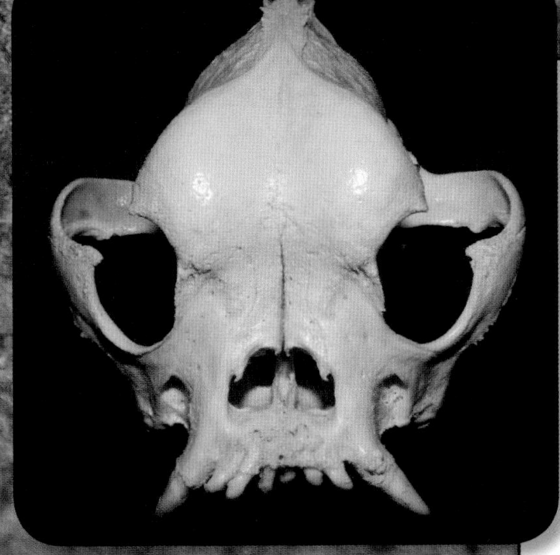

Bulldog Skull

Bulldog

Great Variations in Dog Jaws

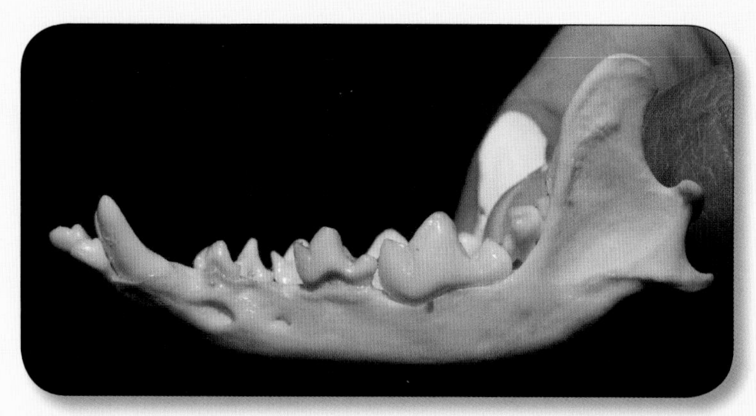

A Chihuahua Jawbone

Chihuahua

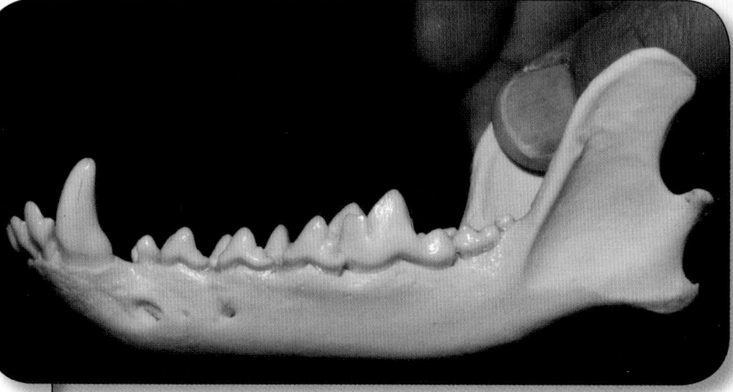

Jack Russell, Terrier Jawbone

Jack Russell, Terrier

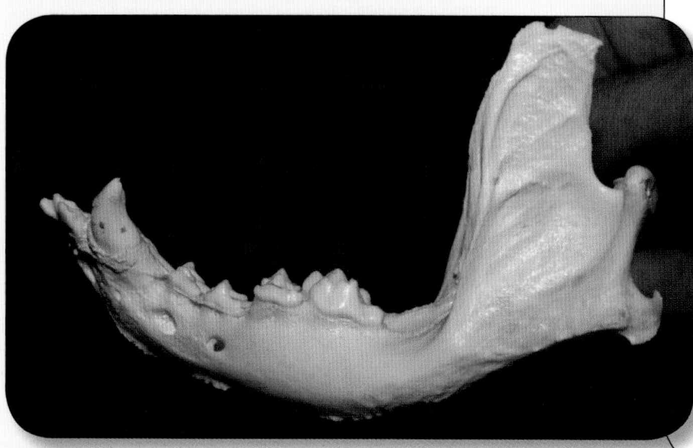

Bulldog Jawbone

Bulldog

Variations in Human Beings (*Homo sapiens*)

The naming of species is subjective and, at times, I find species and genus names unreliable. For example, in the past, evolution scientists thought they could "see" differences between "races" of humans and incorrectly assigned each "race" to a different species based on minor variations, such as skin color and hair texture. White European human beings were assigned these nine species names: *Homo albus,*[2] *Homo alpinus,*[3] *Homo sapiens albus nordicus,*[3] *Homo caucasicus,*[2] *Homo indo-europaeus,*[2] *Homo europaeus,*[3] *Homo indo-europaeus dolichomorphus nordicus,*[3] *Homo indo-europaeus brachymorphus alpinus,*[3] and *Homo nordicus.*[3] Mediterranean people — Italians, Greeks, and Libyans — were assigned to one of these species: *Homo Mediterraneus,*[3] *Homo merridionalis,*[3] or *Homo indo-europaeus dolichomorphus mediterraneus.*[3] People from China, Japan, and Mongolia were named *Homo flavus*[2] or *Homo mongolicus.*[2] People with darker skin were assigned to the species *Homo niger*[2] or *Homo aethiopicus.*[2]

But this was all wrong! Scientists were incorrect in giving human beings different species names based on minor insignificant variations, such as skin color, forehead angle, face shapes, skull shapes, skull size, and the like. Modern genetics has shown that all "races" of humans have similar DNA. Also, all human "races" can interbreed and produce fertile offspring. Actually, biologically there is only one race-the human race. Pygmies, Eskimos, Asians, Blacks, Italians, and Libyans are the same species, namely *Homo sapiens.*

If scientists erred in naming *living* species, which they could study and observe and directly test the ability to produce fertile offspring, it is also possible they might error in naming *fossil* species, which is much more difficult.

Scientists mistakenly thought different "races" of humans represented more than 15 different species.

Variations in American Oysters (*Crassostrea virginica*)

Look at these modern oysters from the Gulf of Mexico. Each shell is different in its size, shape, and appearance, yet all are the same species and can interbreed.[4] What if a scientist was unfamiliar with this type of oyster and found these six shells fossilized miles apart from each other? He or she might incorrectly assign a new species name to each of these six shells.

Just one species of oyster has a tremendous amount of variation in size, shape, and appearance.

Mitten

Three-Finger Mitten

Symmetrical Shape

Variations in Sassafras Leaves
(Sassafras albidum)

Variations also occur in plants. Look closely at the leaves on this page and notice the differences in their shape, veins, and appearance. Despite these drastic differences, all five of these leaves came from *the same tree.* Imagine if a paleontologist was unfamiliar with the sassafras tree and found these five leaves fossilized miles apart from each other. The scientist might incorrectly assign each one to a different genus or species.

Each leaf looks different on one sassafras tree!

Oval

Not Vertically Symmetrical

Comparing Modern and Dinosaur-Era Magnolias

Compare the dinosaur-era fossil leaf on the left with the modern magnolia leaf on the right. To me, the differences between the fossil and the modern leaf are no greater than the variations in the leaves from the same magnolia tree on the next page.

Dinosaur-Era Magnolia
Magnolia magnifolia
New Mexico Museum of Natural History and Science, Albuquerque, USA

Modern Star Magnolia
Magnolia stellata
Missouri Botanical Gardens, St. Louis, USA

From *Evolution*: *The Grand Experiment* video series

"*Magnolias today are in some ways very similar, sometimes considered living fossils of what trees were like during the Cretaceous.*" [5]

— Dr. Thomas Williamson

Dr. Thomas Williamson is curator of paleontology at the New Mexico Museum of Natural History and Science, Albuquerque, New Mexico.

On one magnolia tree, each leaf looks different!

Variations in Star Magnolia Leaves

(Magnolia stellata)

The Naming Game

Let me digress and summarize where we are. ***My experiment predicted I would find modern plant and animal species in dinosaur rock layers if evolution was not valid.*** While Debbie and I photographed many modern-appearing plants and animals found in dinosaur rock layers, not one fossil was assigned a modern species name.

I became plagued with this question: "Why do the fossils look modern but have different species names and many times different genus names assigned to them?" It became more and more evident as our experiment continued that the field paleontologists we interviewed were not going to offer an answer capable of satisfying my curiosity. Since I personally "needed to see" the implied "differences" between the fossil and modern forms that warranted different species names, I decided to compare our photographs of the fossils to living plants and animals. This was by no means an easy task and involved years of work to accomplish. It was during this period that I encountered and became aware of what I now refer to as "The Naming Game."

We have already discussed there is the potential for error in naming a species due to (1) great variations within a single species, and (2) the subjective nature of assigning a genus and species name. My personal experience bore this out in a very tangible way in just the ten years of working on this book. Scientists who support evolution changed genus and species names of some fossils presented here.[6-11] For example, a fossil I photographed at a museum would be one genus and species, only later to be changed to another genus and species. It appears to me that modern scientists frequently disagree on genus and species names of fossils, as well as living organisms.[12] Because of all of this, I have gained a certain amount of "mistrust" when it comes to the scientific naming of a species. Now, as a skeptic, I only trust what I can see with my own eyes, not the assigned names.

Since a picture is worth a thousand words, let's now look at the pictorial evidence I gathered for my experiment and compare fossils to living forms. As you read this book and judge the theory of evolution for yourself, based on what you see, I ask: Is this a case for evolution or a case for doubt?

Author's Note: In the following chapters you will see photos of fossils and photos of living animals and plants organized by groups. Under each photo I will provide the information that will be most helpful. In the first line of each caption will give a short description of the organism. Next you will see the genus name of the organism (in blue), the species name (in red), the rock layer in which the fossil was found (for example, the Triassic, Cretaceous, or Jurassic), and the country where the fossil was found. The last line of each caption tells you where the photo was taken, such as a particular museum, dig site, botanical garden, or zoo. Some of the photographs do not have all of this information because the curators were unable to identify these details.

I purposely left out the ages of these fossils since there is great controversy among scientists concerning how old these fossils actually are. Most scientists who support evolution would suggest that every fossil in this book is more than 65 million years old. Some scientists who oppose evolution suggest that these fossils may be less than 10,000 years old. This is quite a difference of opinion! I will forgo this controversy for the time being and first focus on these living fossils. I will address the age of the fossils in the last book in this series, Volume IV.

The naming of fossil species may erroneously lead us to believe that animals and plants changed over time, but, in reality, it may be only the names that changed.

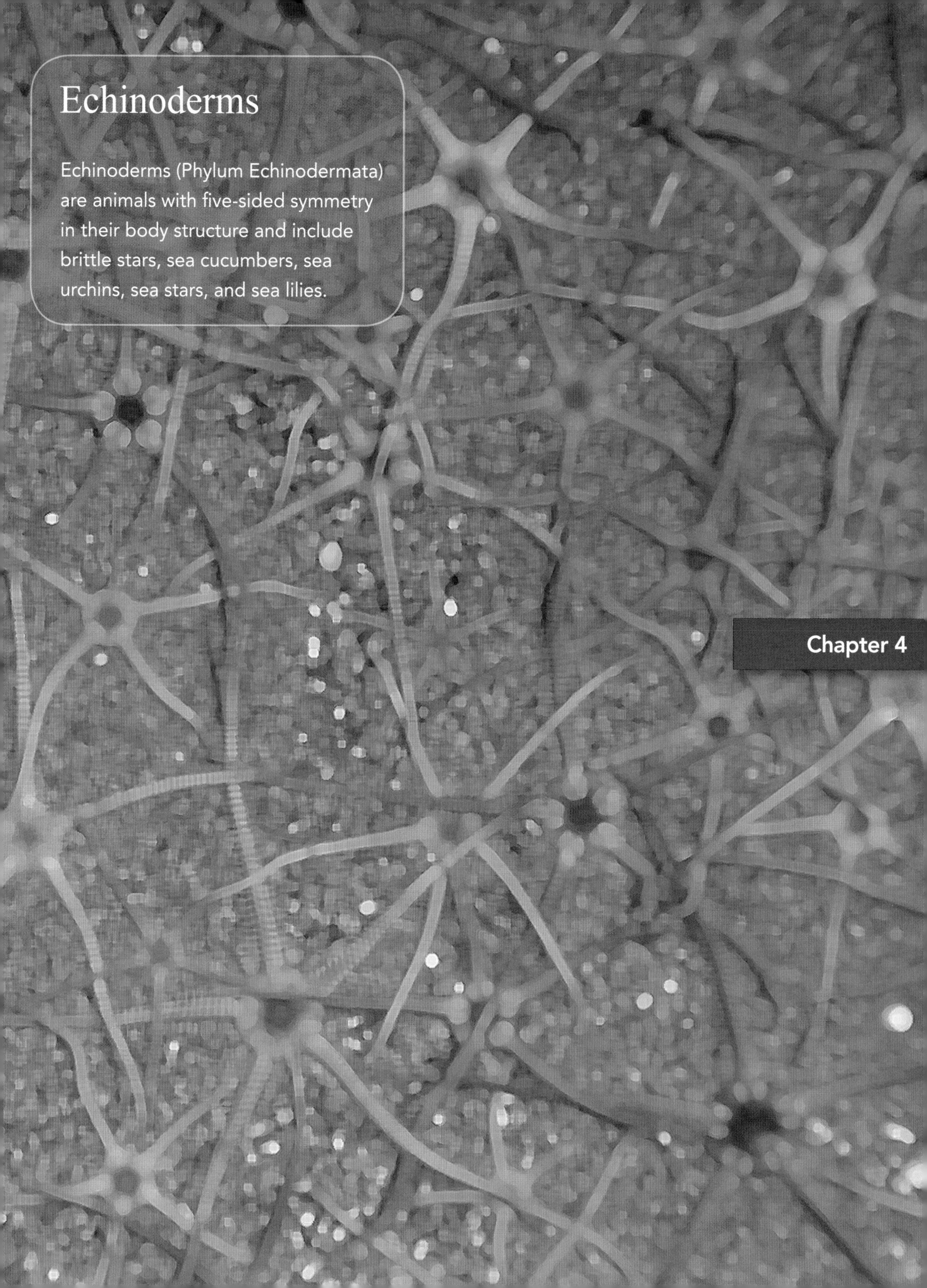

Echinoderms

Echinoderms (Phylum Echinodermata) are animals with five-sided symmetry in their body structure and include brittle stars, sea cucumbers, sea urchins, sea stars, and sea lilies.

Chapter 4

"Even at the first dinosaur dig site we visited, we ran across a modern-appearing brittle star."

The Wyoming Dinosaur Center, Thermopolis

On our first outing in 1997, while traveling in our rental RV called "Betsy," we photographed fossils at the Wyoming Dinosaur Center in Thermopolis. This active dinosaur dig site (left), containing thousands of dinosaur bones from the Jurassic rock layers is colocated with a museum. The museum houses not only the Jurassic fossils found here but other fossils from around the world. It was here that I ran across a fossil labeled *Ophiopetra* (next page). At the time, I did not recognize its name or significance. I thought this fossil was an extinct invertebrate from the dinosaur era.

Later, I placed a photograph of a modern brittle star next to this fossil (next page). To me, it appears this form of life has not changed significantly since the time of the dinosaurs. Would you agree?

One further note: Notice that *every* fossil in this chapter was assigned a completely different genus name (in blue) than the living form, even though they look nearly the same.

Brittle Stars Found in Dinosaur Layers

Dinosaur-Era Brittle Star
Ophiopetra (species undetermined)
Jurassic, Kelheim, Germany
Wyoming Dinosaur Center, Thermopolis

Living Brittle Stars
Ophiopholis (Species undetermined)
Sea of Alaska[1]

Later, we photographed two other brittle stars in Australia which once again appear similar.

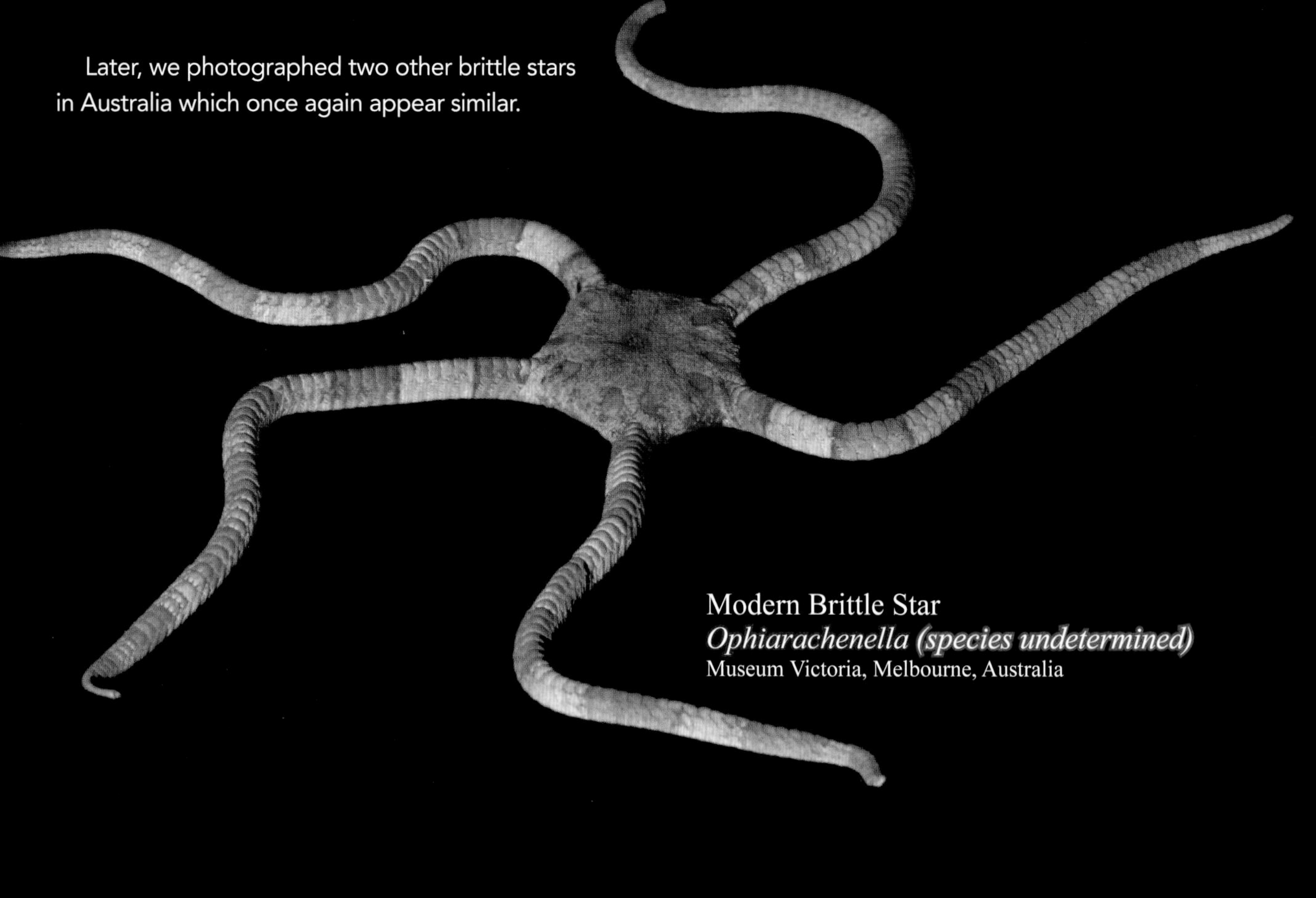

Modern Brittle Star
Ophiarachenella (species undetermined)
Museum Victoria, Melbourne, Australia

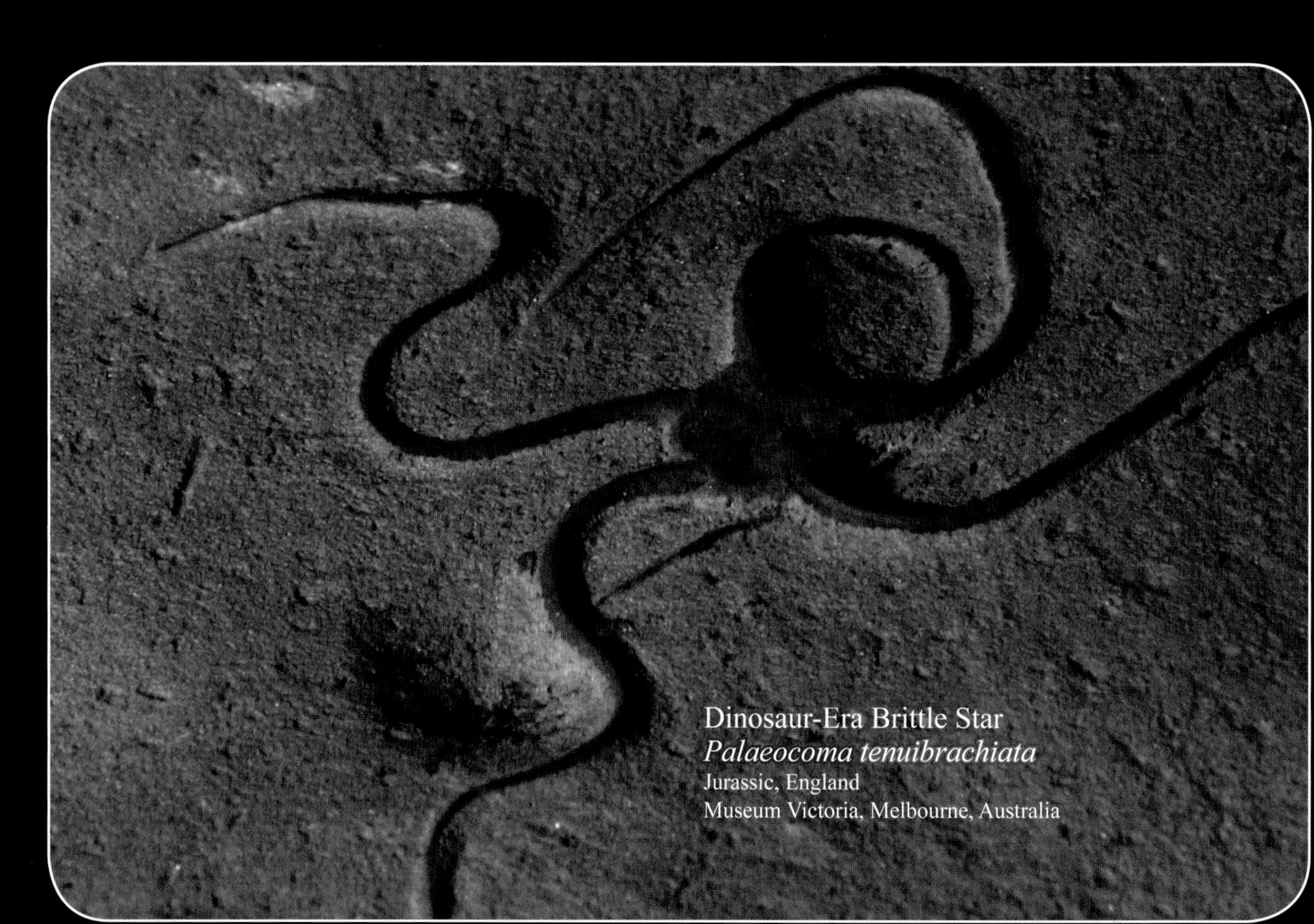

Dinosaur-Era Brittle Star
Palaeocoma tenuibrachiata
Jurassic, England
Museum Victoria, Melbourne, Australia

Dinosaur
Diplodocus hayi
Jurassic, Wyoming
Houston Museum of
Natural Science, Texas, USA

Right: *This dinosaur, Diplodocus, was alive at the same time as the fossil brittle star on the previous page.*

Sea Urchins

Most sea urchins look like a tennis ball with spikes sticking out of it. Look at this *fossil* sea urchin (top of next page) that was found in dinosaur rock layers in England. During the fossil preparation, museum curators physically separated the round body from the spines. Try to imagine what this sea urchin looked like when it was alive, with its spines attached to the body. Now carefully compare this fossil to the living sea urchin below. Look at the shapes of the spines, the bodies, and the soft tissue ridges. I ask this simple question: Are the differences between these two sea urchins, one living and one fossilized, greater than the differences between a Chihuahua and a Boxer? (Remember the huge variations in the skulls, jaws, legs, and teeth within the same species — *Canis familiaris*.)

Living Pencil Sea Urchin
Eucidaris tribuloides
St. Thomas, USVI

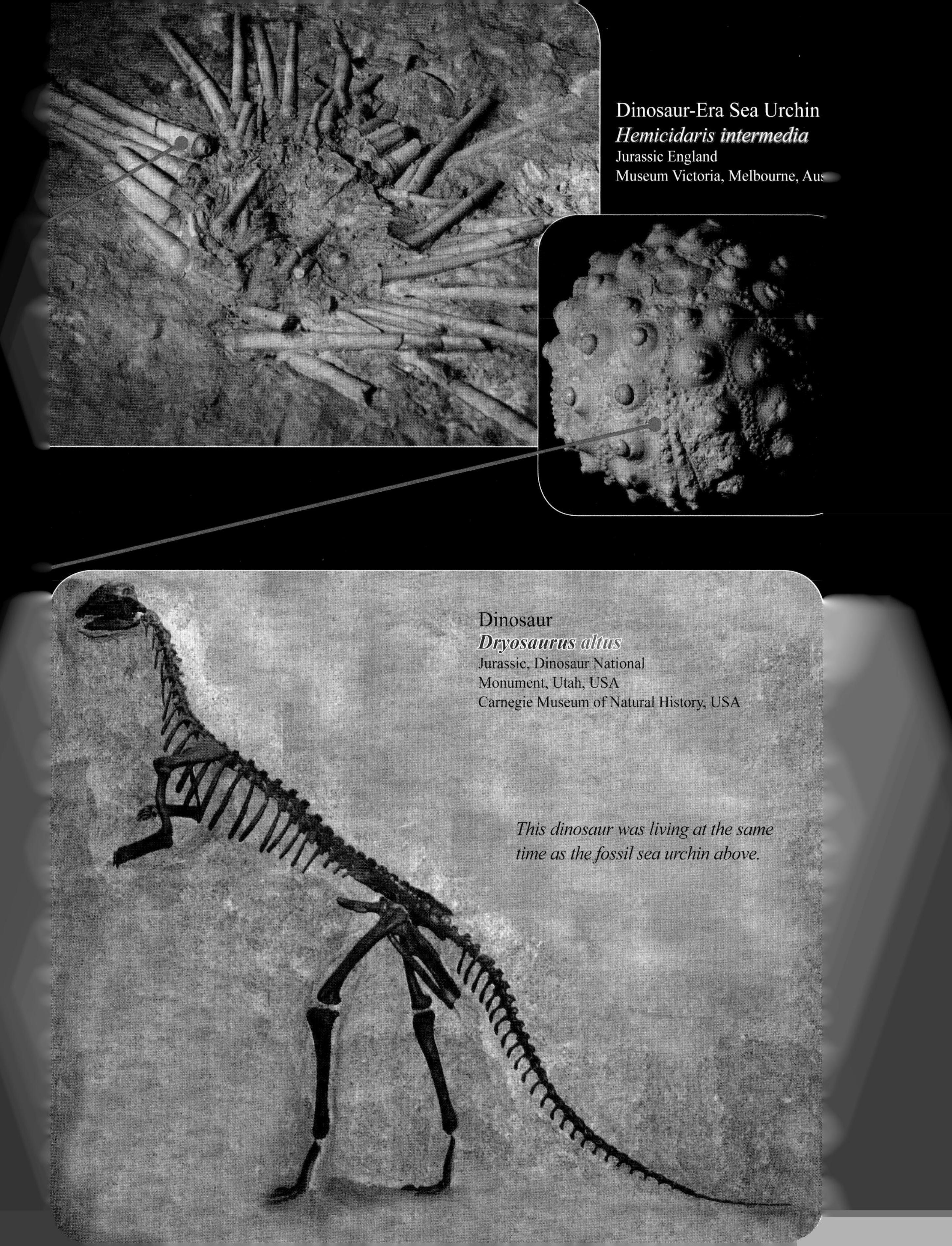
Dinosaur-Era Sea Urchin
Hemicidaris intermedia
Jurassic England
Museum Victoria, Melbourne, Aus
Dinosaur
Dryosaurus altus
Jurassic, Dinosaur National
Monument, Utah, USA
Carnegie Museum of Natural History, USA
This dinosaur was living at the same time as the fossil sea urchin above.

Comparing Modern and Dinosaur-Era Sea Biscuits

One peculiar variety of sea urchin, called a sea biscuit, does not have spines and, as the name implies, looks like a biscuit. One day, while roaming the halls of the Carnegie Museum in Pittsburgh, I photographed a fossil sea biscuit (right) that reminded me of the living purple heart sea biscuit (below). From my perspective, they look similar. What do you think?

"Extinct" Dinosaur-Era Sea Biscuit
Holaster (species undetermined)
Cretaceous, Dover, England
Carnegie Museum of Natural History, USA

Living Purple Heart Sea Biscuit
Spatangus purpureus
Ireland

Dinosaur-Era Starfish

Now look at this fossil sea star and this modern sea star. The distinct genus names (in blue) imply significant differences exist between them.

Dinosaur-Era Sea Star
Asteroidea (species undetermined)
Cretaceous, South Australia
South Australian Museum, Adelaide

Living Sea Star
Asterias laevis
South Australian Museum, Adelaide

A Night Dive That Almost Ended the Challenge

One moonless night, I found myself sitting in a boat in the Caribbean Sea off the coast of Cozumel, Mexico, trying to win this challenge. I was about to go on my first *night* scuba dive. It seemed like a nice idea when I signed up 12 hours earlier in the bright morning sun, but now I was beginning to have second thoughts. The more I thought about it, the more I was plagued with the idea of being bitten by a shark or drowning at night. To be attacked by a shark in the daytime would be bad enough, but the thought of a shark attack in the dark night, well, that seemed a bit much.

My jitters were multiplying by the minute as our group of divers prepared to get into the water. After entering the water, the better part of my senses *finally* took hold of me and I told the dive master that I had changed my mind. I wanted to reverse course, get back into the boat, and wait for the group topside. He told me in broken English "*You will be fine, Carl. You will be fine! No problema! I will stay with you. No problema!*"

Due to his persistence, he talked me into diving. I found myself descending into the dark, black water, and soon it became all too clear how foolish this expedition was. (To this day, I wish I had followed my instincts and just gotten back into the boat.) Later, during that dive, I nearly ran out of air taking a photograph of a sea cucumber (bottom of next page). After taking the photograph, I realized the dive master and the entire group of divers left me alone in the dark 50-foot deep water! At this point, wisdom (outflanked by fear) told me it was time to get out of the abyss — and fast.

I hoped my air would not run out as I ascended to the surface, all alone, and prayed that a shark was not cruising toward my legs. I later wondered why I had put myself in such danger just for a photograph of a sea cucumber! My friends and family have grown accustomed to me going to great lengths to get photos of ordinary things. It's just a "Carl thing."

Sea cucumbers (next page) look like bananas or, as the name suggests, cucumbers slowly moving across the ocean floor. I have seen many species of these creatures when diving at the Great Barrier Reef in Australia, the Virgin Islands, and Cozumel, Mexico. Externally, sea cucumbers do not appear to have five-fold symmetry like a sea star, but when you dissect them, you can see it. Because of this, they are in the same group of animals as sea stars (echinoderms).

Compare the photograph I took that night of a living sea cucumber (on the next page) to a fossil sea cucumber found below a dinosaur fossil layer.* Ignoring colors, which are not preserved in fossils, do they look fundamentally unchanged?

Fossil Sea Cucumber
Achistrum (species undetermined)
Pennsylvanian,* Essex, Illinois, USA
Humboldt State University, USA

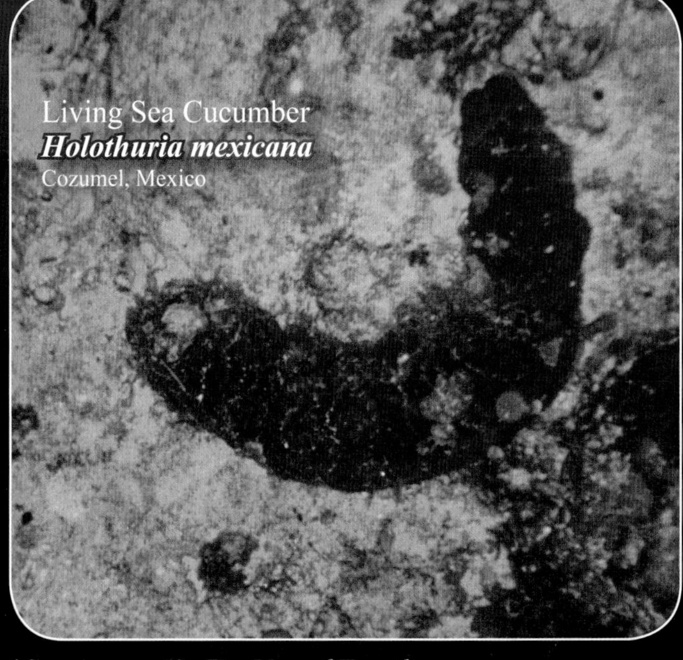

Living Sea Cucumber
Holothuria mexicana
Cozumel, Mexico

***See Appendix B:** *Use of Fossils Found below Dinosaur Layers*

Sea Cucumbers

Dinosaur-Era Crinoid Ring
Crinoidea (species unspecified)
Triassic, Arizona, USA
Petrified Forest National Park, USA

Long-Stalked Crinoids

Crinoids, also called sea lilies, are the fifth and last major group of echinoderms living today. They have a five-fold symmetry in their feeding arms; that is, the number of feeding arms is either 5 or 10 or 15, and so forth.

Near my childhood home, you can stop at almost any bluff along the highway and find dozens of sea lily rings, also called crinoid rings, like the one on the left. These individual rings form the long stalks of stemmed sea lilies. I had always thought these crinoid rings were evidence of change or evolution. I thought they were from an extinct strange animal. But on my expedition, I was able to compare a fossil crinoid to a living crinoid (on the next page) and I now have my doubts.

Above: *This dinosaur, Psittacosaurus, was found in the early Cretaceous fossil layers, the same layers in which the fossil crinoids on the next page were found.*

Dinosaur-Era Stemmed Crinoid
Eucrinus liliiformis
Triassic, Germany
Museum Victoria, Melbourne, Australia

Comparing Fossil and Modern Stalked Crinoids

Stemmed Crinoid from Dinosaur Layer
Isocrinus australis
Cretaceous, South Australia
South Australian Museum, Adelaide

Modern Stemmed Crinoid
Metacrinus cyaneus
Continental Shelf, South Australia
South Australian Museum, Adelaide

Feather Stars

In addition to the long-stemmed stalked crinoids, there are other types of sea lilies living today without stalks, called feather stars. One day while we were filming at the Alaska Sea Life Center, I saw a living feather star and realized I had found a match to a fossil collected nearly 5,000 miles away in Germany. To me, these two invertebrates look similar.

Dinosaur-Age Feather Star
Pterocoma pennata
Jurassic, Germany
California Academy of Sciences, USA

Living Feather Star
Florometra serratissima
Alaska Sea Life Center, USA

Living Feather Star
Oxycomanthus bennetti
Wakatobi, Indonesia

Summary

So concludes this chapter on animals with five-fold symmetry. In order to ensure that my point does not go unnoticed, let me reiterate: Examples of *all* five major classes of echinoderms living today [2] — starfish, brittle stars, sea urchins, sea cucumbers, and sea lilies — have been found in dinosaur rock layers. Even though these fossils look very similar to the modern varieties, they have been assigned completely different genus and species names! If I ignore the names, it appears that evolution has not occurred. You may be asking yourself if the same can be done for other groups of animals and plants, such as crustaceans, shellfish, corals, birds, amphibians, reptiles, mammals, and flowering trees? Hang on to your seats!

Above: *Long-necked dinosaurs were found in the same layer (Jurassic) as the brittle star, sea urchin and the feather star.*

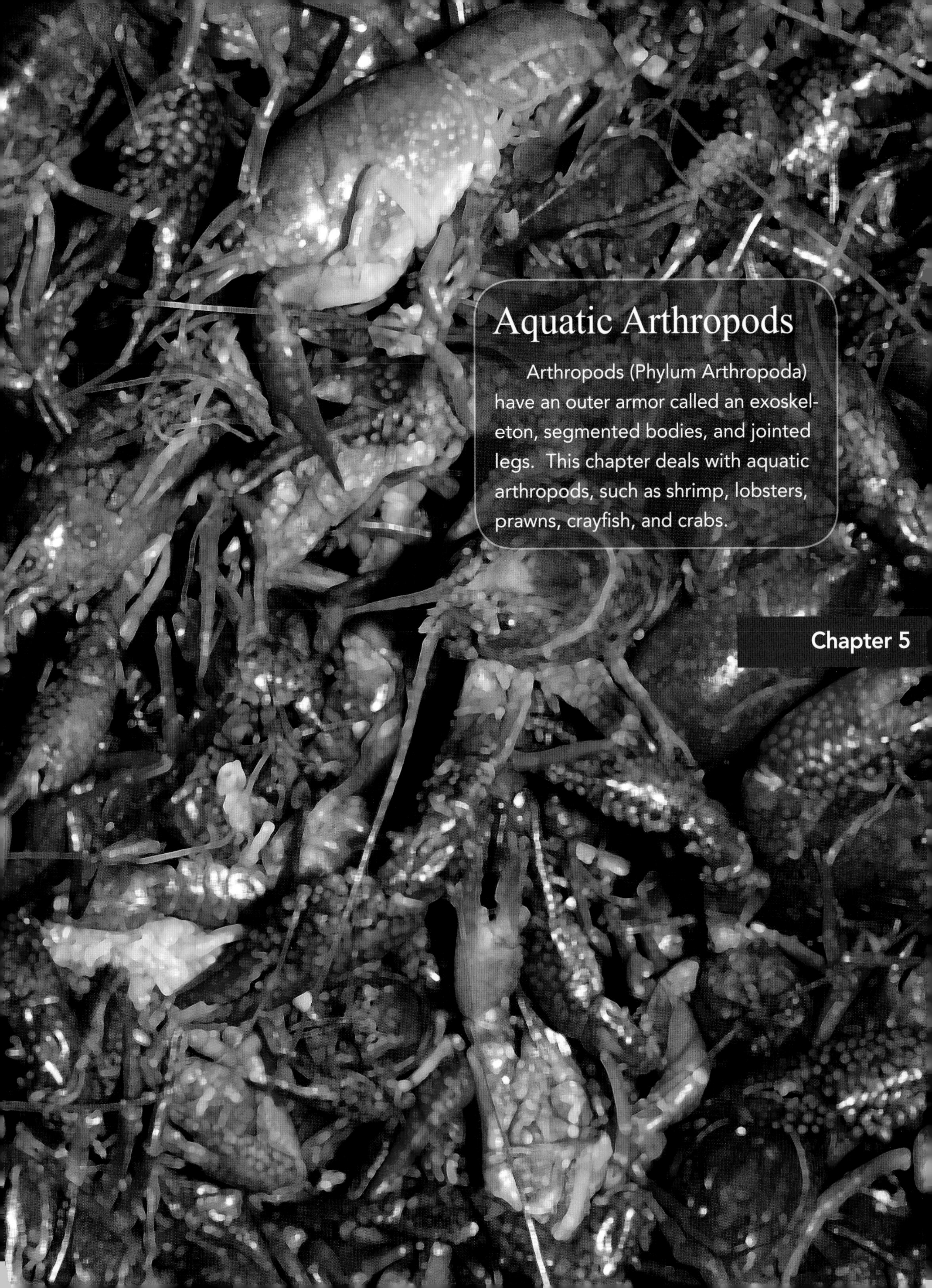

Aquatic Arthropods

Arthropods (Phylum Arthropoda) have an outer armor called an exoskeleton, segmented bodies, and jointed legs. This chapter deals with aquatic arthropods, such as shrimp, lobsters, prawns, crayfish, and crabs.

Chapter 5

"In Germany, I came across a fossil that made my eyes light up."

Statue in Eichstatt, Germany

Debbie in Eichstatt

Jura Museum in Eichstatt, Home of Archaeopteryx

The Solnhofen Tile Quarry: Home of *Archaeopteryx*

During the third year of our adventure, in 2000, Debbie and I traveled to Europe. A portion of our trip was dedicated to photographing fossils at the world-famous Solnhofen Fossil Quarry in Germany (next page). Over the last 150 years, many important fossils have been found here, including the famous toothed bird *Archaeopteryx* (next page, once thought to be an intermediate step between birds and dinosaurs), the dinosaur *Compsognathus* (next page), as well as tens of thousands of other fossils. This quarry is still used to produce stone tiles that are shipped all over the world.

Many of the fossils from this quarry are housed in the Jura Museum in Eichstatt, Germany, just a few miles away from the quarry, and because of this, the town names of Solnhofen and Eichstatt have become synonymous with the fossils from this region. It was here where I came across a fossil that definitely made my eyes light up.

Solnhofen Quarry

Quarry worker preparing tiles

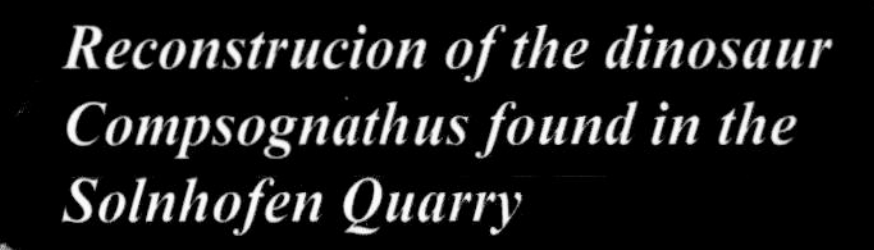

Reconstrucion of the dinosaur Compsognathus found in the Solnhofen Quarry

Archaeopteryx found in Solnhofen Quarry

Dinosaur-Era Shrimp
Antrimpos speciosus
Jurassic, Solnhofen, Germany
Jura Museum, Germany

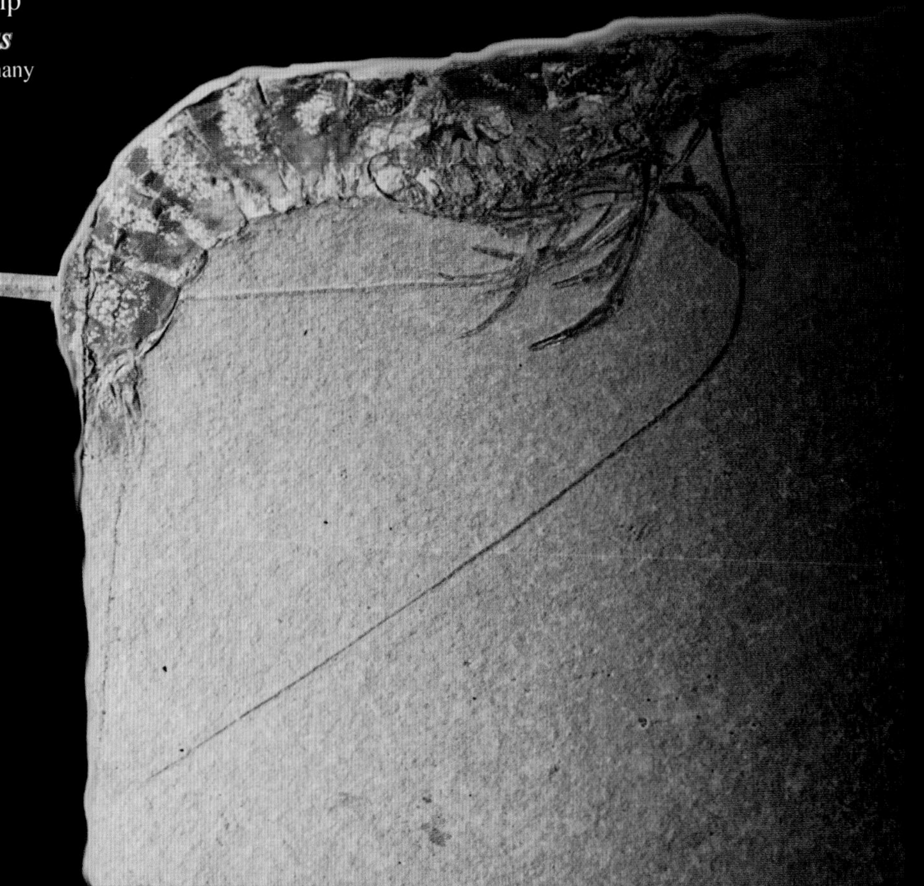

Shrimp

At the Jura Museum in Germany, just a few feet from the original Archaeopteryx fossil, I saw a fossil shrimp. It reminded me of the gulf shrimp I had eaten in Louisiana. Was this shrimp from dinosaur times the same shrimp I love to eat? Looking at the fossil photo Debbie took at the museum and the photo of the live shrimp, I think they are extremely similar. Do you agree with this assessment?

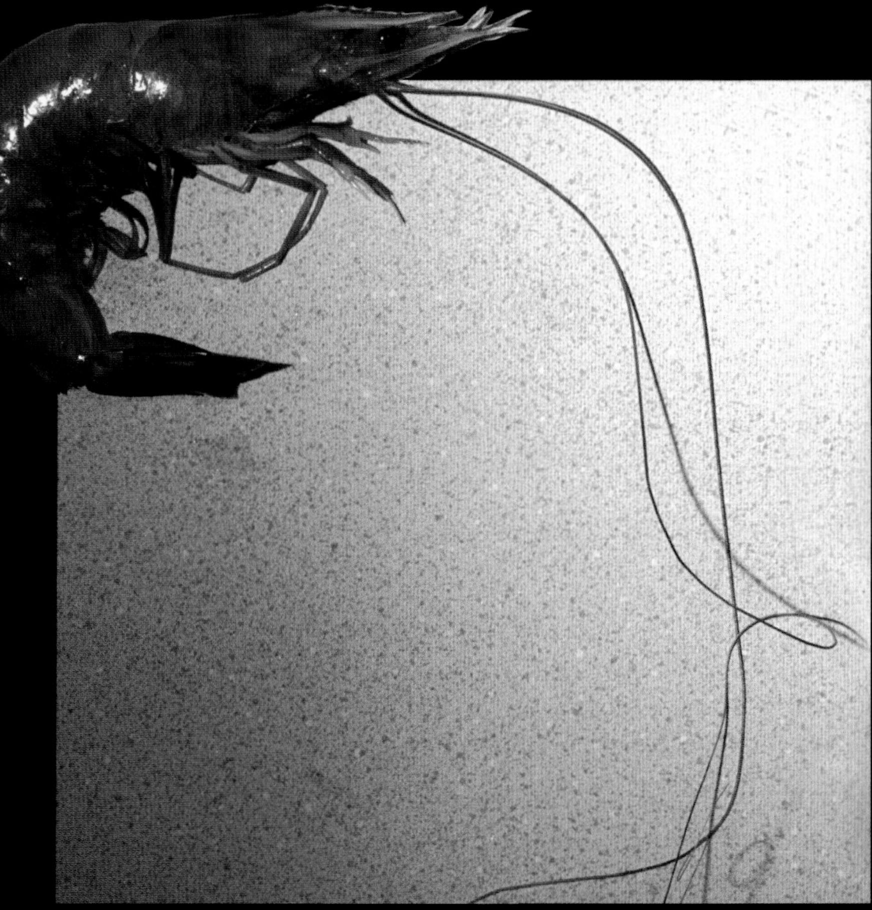

Modern Gulf Shrimp
Tail rolled up
Litopenaeus setiferus
Louisiana, USA[1]

Lobsters Found with *Archaeopteryx*

Now compare this modern Maine lobster with its large claws to this fossil lobster found at the Solnhofen Quarry.

Live Maine Lobster
Tail spread
Homarus americanus
USA

Dinosaur-Era Lobster
Tail narrowed
Eryma leptodactylina
Jurassic, Solnhofen, Germany
Chicago Field Museum, USA

Spiny Lobster at Solnhofen

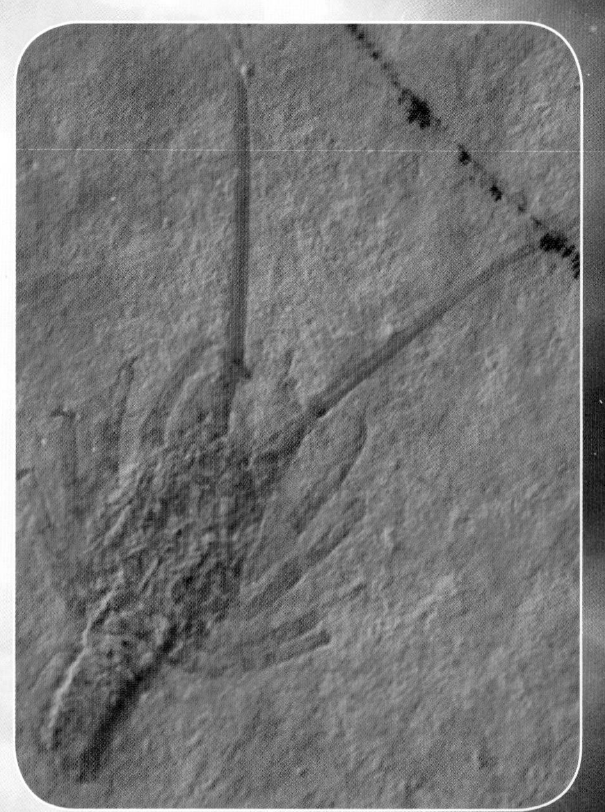

Juvenile Dinosaur-Era Spiny Lobster
Antennae pointed forward
Palinurina longipes
Jurassic, Solnhofen, Germany
Carnegie Museum of Natural History, USA

Spiny Lobster
Antennae pointed backwards
Panulirus penicillatus
Hawaii, USA

Spiny lobsters do not have the succulent large claws of a Maine lobster, hence they are not sought after as much for food. I have photographed spiny lobsters in the Caribbean and on the Great Barrier Reef. Compare this live spiny lobster above to the fossil lobster from the Solnhofen Quarry.

Horseshoe Crabs

Dinosaur-Era Horseshoe Crab
Mesolimulus walchi
Jurassic, Solnhofen, Germany
Jura Museum, Germany

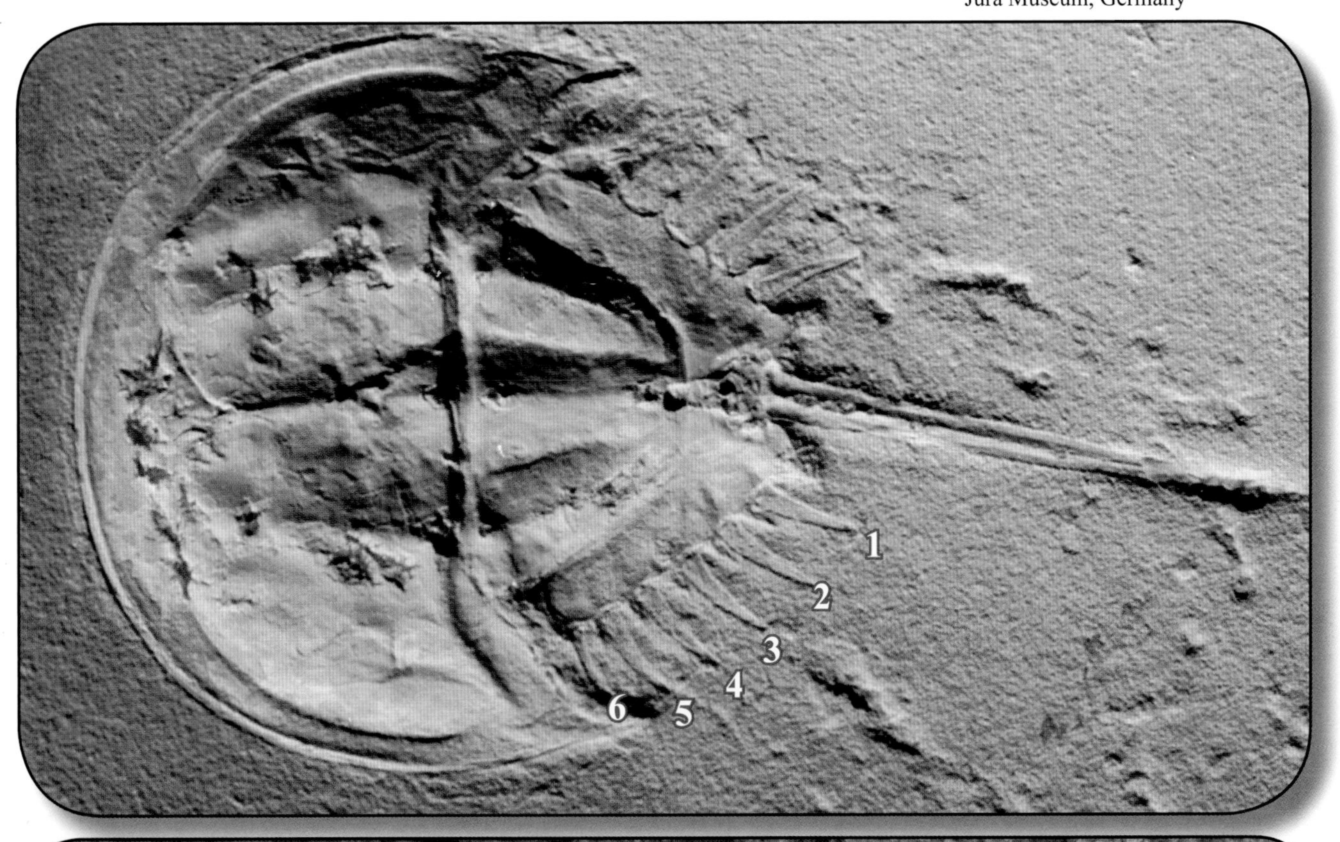

Horseshoe Crab
Limulus polyphemus
World Aquarium, Missouri, USA

When you look at the photos of these horseshoe crabs (with different genus and species names), you may marvel that modern-appearing horseshoe crabs lived with dinosaurs.

Horseshoe Crabs — Underneath View

Dinosaur-Era Horseshoe Crab
Underneath surface
Mesolimulus walchi
Jurassic, Solnhofen, Germany
Jura Museum, Germany

See Appendix A:
The Influence of Fossil Orientation

Horseshoe Crab
Underneath surface
Limulus polyphemus
World Aquarium, Missouri, USA

Prawns

Now look at this unusual fossil from Solnhofen with its extremely long claws and compare it to this living freshwater prawn.

Dinosaur-Era Fossil
Mecochirus longimanus
Jurassic, Solnhofen, Germany
Wyoming Dinosaur Center, Thermopolis, USA

Freshwater Prawn
Macrobrachrium rosenbergii
World Aquarium, Missouri, USA [2]

Living Crayfish
Procambrus clarkii
Louisiana, USA[3]

Dinosaur-Era Crayfish
Eryma leptodactylina
Jurassic, Germany
California Academy of Sciences, USA

Crayfish Found in Jurassic Rock Layers

The *live* crayfish (above) has its tail rolled up, its antennae pointed *forward*, and its left claw *open*. The *fossil* crayfish (right) has its tail less tucked up, its left claw *closed*, and its antennae pointing *backward*. If you ignore these particular body positions and color, which is not preserved in fossils, they look very similar.

Dinosaur-Era Crab
Genus Undetermined
Cretaceous, Tennessee
Chicago Field Museum, USA

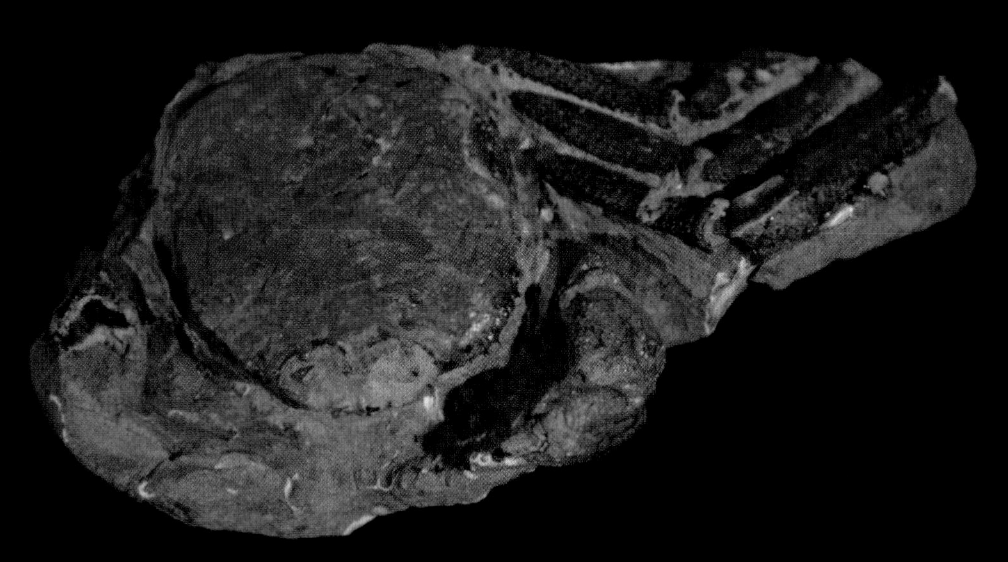

Crabs

The fossil crab above is the only fossil in this chapter not found at Solnhofen. It was found in a dinosaur layer in Tennessee. Compare it to this modern crab below. These two crabs were photographed at different angles. If I could photograph both at the same angle, I believe an even better match could be made. At the very least, I conclude that crabs were alive at the time of the dinosaurs, and they looked very similar.

Living Crab
Unidentified
Kona, Hawaii, USA

See Appendix A:
The Influence of
Fossil Orientation

Summary

All of the major aquatic arthropod groups living today — shrimp, crayfish, fresh water prawns, lobsters, crabs, and even horseshoe crabs [4,5] — have been found in dinosaur fossil layers and appear amazingly similar to modern forms.

Land Arthropods

Land arthropods (Phylum Arthropoda) include familiar creatures, such as insects, spiders, scorpions, millipedes, and centipedes. This group, like aquatic arthropods, has a segmented body, an exoskeleton and jointed legs.

"The last thing I was expecting to see at Petrified Forest was a termite nest."

Petrified Forest National Park

Coelophysis Dinosaur

Petrified Forest National Park, Arizona

Petrified Forest National Park in Arizona is most famous for the thousands of fossilized trees spread over 93,533 acres. Few visitors are aware that two types of dinosaurs (*Coelophysis* and *Chindesaurus*) have been found here too.

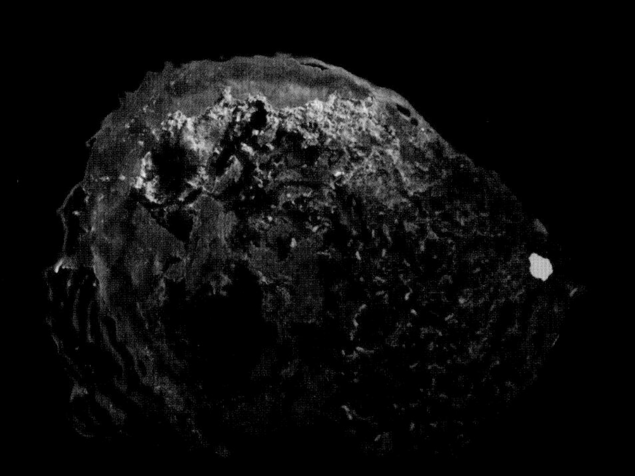

Modern Termite Nest
Coptotermes acinaciformis
Tamworth, NSW Australia

Termite Nest Found at Petrified Forest

After Debbie photographed the petrified trees in the park, we stopped in at the Painted Desert Visitor Center to photograph the dinosaur bones and dinosaur models. Here in a display case was a fossil termite nest that was found at the park. I mused when I compared this fossil (below, left) to the museum drawing of the inside of a modern termite nest (below, right) and a photograph of a modern termite nest (above).

Dinosaur-Era Termite Nest
Triassic, Arizona, USA

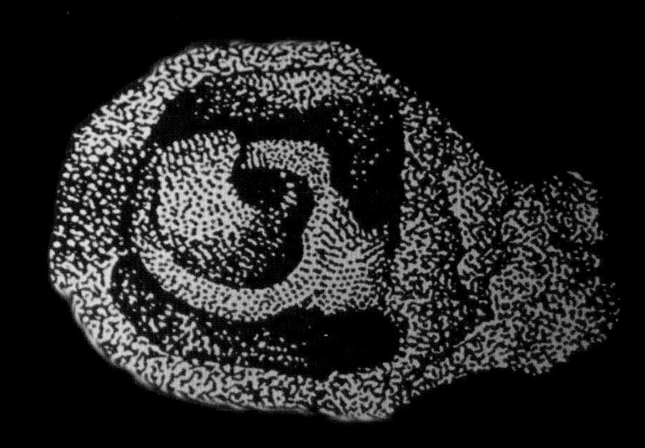

Museum Drawing Modern Termite Nest
Petrified Forest National Park Visitor Center, USA

Other Fossil Insects

Author's Note: The fossils on these next six pages were found at the Solnhofen Quarry in Germany described in Chapter 5.

Dinosaur-Era Dragonfly
Urogomphus giganteus
Jurassic, Solnhofen, Germany
Carnegie Museum of Natural History, USA

Dragonflies

When I started my expedition, I was aware that dragonflies had been found in dinosaur rock layers, but these fossils were assigned a different genus — not closely related to modern dragonflies. I always believed this; yet, when I put the fossil dragonfly next to a living dragonfly, I was not so convinced.

Dragonfly (Blue Dasher)
Pachydiplax longipennis
Louisiana, USA [1]

Fork-tailed Bush Katydid
Scudderia furcata
Missouri, USA[2]

"Extinct" Dinosaur-Era Insect
Pycnophlebia speciosa
Jurassic, Solnhofen, Germany
Jura Natural History Museum, Germany

Katydids with Dinosaurs?

Compare this living katydid to this dinosaur-era fossil.

Notice the legs and the antennae are in different positions.

Water Skaters

Water skaters are spider-like insects that scoot on the surface of the water in streams, ponds, and lakes. They have many names. People also refer to them as water striders, pond skaters, or water skeeters. Compare this living water skater, sitting on the back of my hand (below), to this fossil from Solnhofen (right). I don't consider this an exact match because the front legs on the living form are shorter than the legs of the fossilized water skater. But if I take into consideration the great variability in the length of legs in other species of animals, I have to conclude that dinosaur-era water skaters were similar to those of today.

Water Skater
Gerris remigis
Missouri, USA[3]

Dinosaur-Era Water Skater
Chresmoda obscura
Jurassic, Solnhofen, Germany
Jura Natural History Museum, Germany

Waterbug

Waterbug
Lethocerus grandis
Jura Natural History Museum, Germany

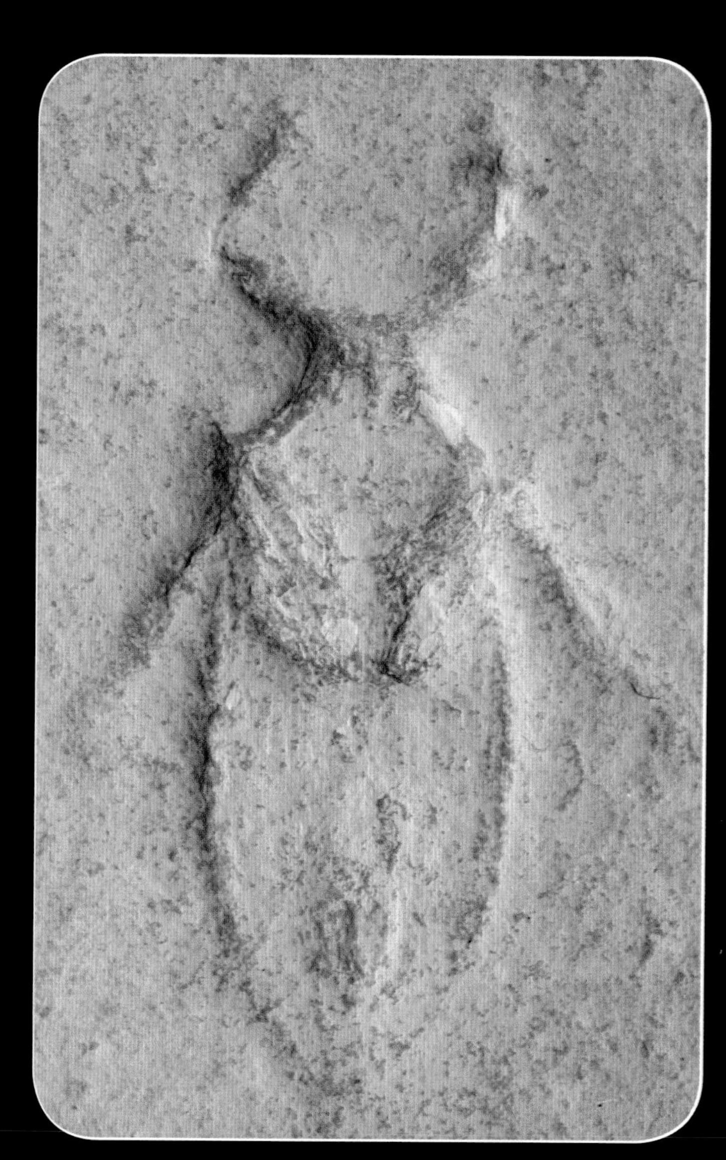

Dinosaur-Era Waterbug
Mesobelostomum deperditum
Jurassic, Solnhofen, Germany
Jura Natural History Museum, Germany

Woodwasps and Beetles

Modern Woodwasp
Urocerus gigas
Jura Natural History Museum, Germany

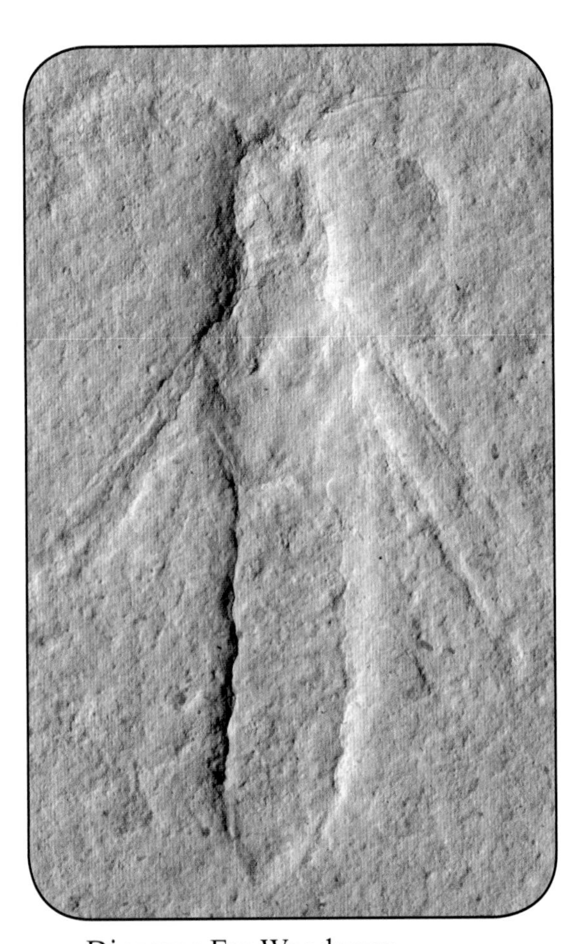

Dinosaur-Era Woodwasp
Myrmicium (species unidentified)
Jurassic, Solnhofen, Germany
Jura Natural History Museum, Germany

Dinosaur-Era Beetle
Wings extended
Jurassic, Solnhofen, Germany
Jura Natural History Museum, Germany

Modern Beetle
Wings at rest
Rhizotrogus aestivus
Jura Natural History Museum, Germany

Scorpionfly

Dinosaur-Era Scorpionfly
Orthophlebia lithographica
Jurassic, Solnhofen, Germany
Jura Natural History Museum, Germany

Modern Scorpionfly
Panorpa communis
Jurassic, Solnhofen, Germany
Jura Natural History Museum, Germany

Modern Scorpionfly
Natural position
Unidentified species
Sussex, England

Mayflies

The next fossil is a mayfly larvae (top of opposite page) found in Australia. Few are familiar with or have ever seen a mayfly larvae. When I located a photo of one and placed it next to this fossil, it appeared very similar; both have the three-tail cerci, the segmented body, and the forward-placed legs.

Mayfly
Scotland, UK

Dinosaur-Era Mayfly Larvae
Australurus plexus
Cretaceous, Australia
Museum Victoria, Melbourne, Australia[4]

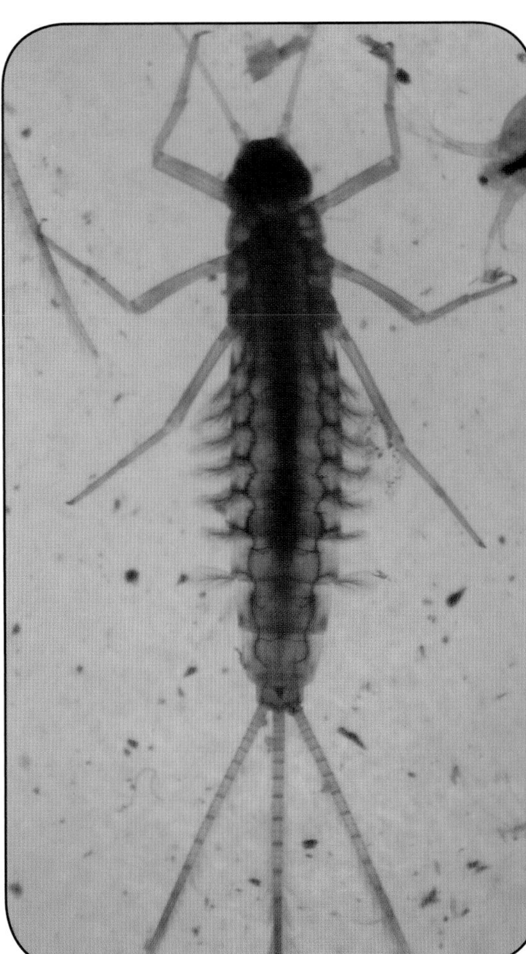

Living Mayfly Larvae
Ipswich, UK

Crickets Living at the Same Time as *Deinonychus*

Since I have fished with crickets for years, I am very familiar with cricket anatomy. These insects have the all too familiar large back legs, small front legs, front antennae, and the four rear antennae (cerci). Now compare this fossil cricket to the modern form. To me, these look like the same species. (Unfortunately, the museum was not able to provide a genus or species name.)

Dinosaur *Deinonychus* Eating Juvenile Dinosaur *Tenontosaurus*
Cretaceous, Montana, USA
Sam Noble Oklahoma Museum of Science and Natural History, USA

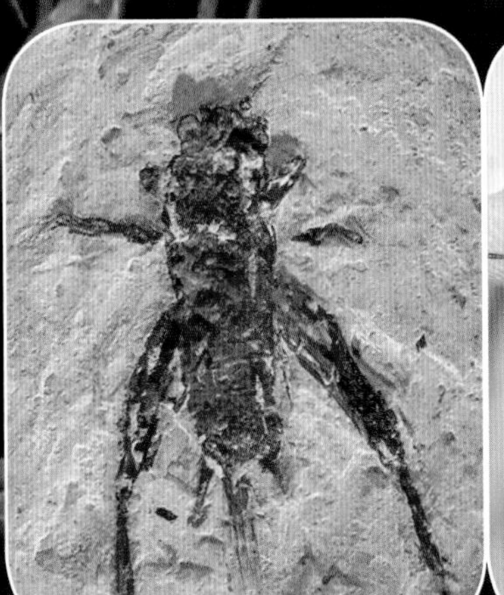

Dinosaur-Era Cricket
Cretaceous, Brazil
Mesalands Community College Natural History Museum, New Mexico, USA

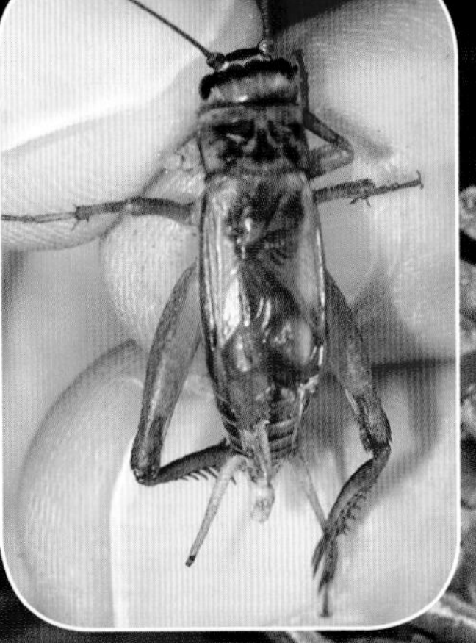

Common House Cricket
Acheta domesticus
Missouri, USA[3]

Cockroaches, Butterflies, and Bees

With each interview and each museum we visited, my list of modern types of insects continued to grow. In fact, I found examples of *every* major insect group (order) living today. [6]

Dr. Clemens: "*The insects go through* [the event that killed off the dinosaurs] *with apparently little change...We're talking about more than cockroaches. We're talking about butterflies and bees.*" [7]

Dr. William Clemens, professor of integrated biology, University of California, Berkeley.

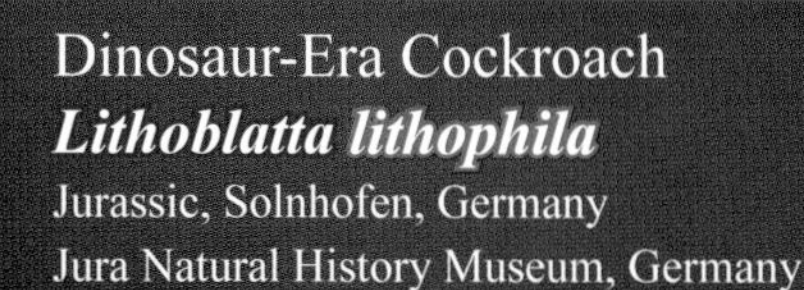

Dinosaur-Era Cockroach
Lithoblatta lithophila
Jurassic, Solnhofen, Germany
Jura Natural History Museum, Germany

Centipedes and Millipedes, Dinosaurs and Pterosaurs

Above: *Flying reptiles (pterosaurs) and dinosaurs are commonly found with modern-appearing insects.*

Daddy Long Leg

Dr. John Long, Paleontologist and Head of Science, Museum Victoria, Melbourne, Australia.

Mite

Dr. Long: [During the time of the dinosaurs]..."*there were insects of modern aspect, as well as spiders, mites, scorpions, millipedes, centipedes, and most things you would find crawling under rocks or soil today.*" [8]

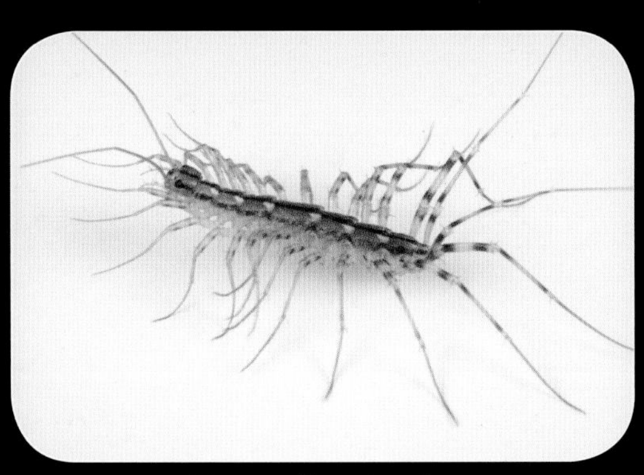

House Centipede

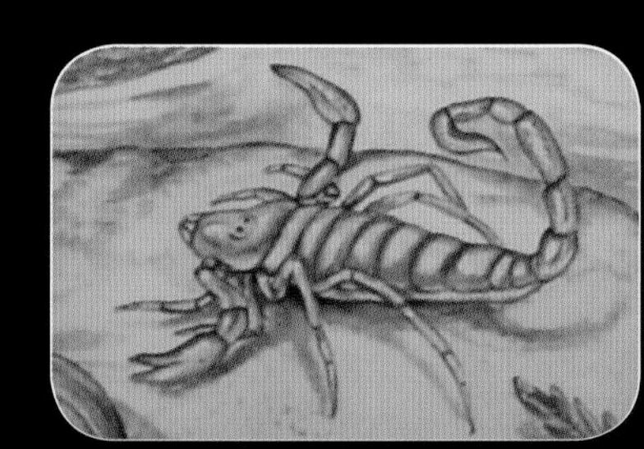

Dinosaur-Era Scorpion
Sam Noble Oklahoma Museum of Science and Natural History, USA

Summary

In dinosaur rock layers, I found examples of all of the major insect orders living today [7]; both major Myriapod classes living today (centipedes and millipedes) [8, 9, 10]; and all of the major Arachnid orders living today.[11] Amazing.

Millipede

Bivalve Shellfish

This chapter deals with the most common class of shellfish living today, bivalves (Phylum Mollusca). As the name implies, these shellfish have two symmetrical shells and include such familiar varieties as scallops, oysters, clams, and mussels.

Chapter 7

"The most famous shellfish living today, the scallop, was also alive during the time of the dinosaurs."

Dinosaur-Age Scallop

Most everyone is familiar with scallops. They are the icon for the Shell Oil Company. A scallop shell has two main parts: the shell with its "ribs," and the hinge joint with its "ears." The delicate "ears" of the hinge joint tend to break off easily, as seen on the previous page. Compare a living scallop (above, right) with this dinosaur-era scallop (right). Except for the fact that the "ears" and the left edges (x) of ribs 15, 16, and 18 are missing, they look similar. Even so, according to the science of evolution, the fossil's name tells us this particular sea creature has gone extinct. Do you agree?

Modern Scallop
Delicate "ears" attached
Aequipecten muscosus
Florida, USA

Dinosaur-Era Scallop
Delicate "ears" missing
Pseudopecten aequivalis
Jurassic, England
Warwickshire Museum, England

Saltwater Clam Found at Dinosaur Provincial Park

Saltwater clams are another familiar variety of bivalve shellfish. The clam below, found at Dinosaur Provincial Park in Alberta, Canada, looks like an ordinary saltwater clam living today. These two shellfish look like they could reproduce. Wouldn't you agree?

Dinosaur-Era Clam
Artica ovata
Cretaceous, Alberta, Canada
Dinosaur Provincial Park, Canada

Modern Clam
Mercenaria mecenaria
Virginia, USA

Freshwater Clams

Compare this freshwater clam with a clam found at Petrified Forest National Park, along with the dinosaur *Coelophysis*. Taking the fossilization process into account, these two clams appear similar.

Modern Freshwater Clam
Unio mancus
Spain

Dinosaur-Era Freshwater Clam
Elliptio (species undetermined)
Triassic, Petrified Forest National Park
Arizona, USA[1]

Mussels

Mussels are the famous bivalve shellfish served in fine restaurants, especially in Brussels. (One should eat "mussels in Brussels" if ever afforded the opportunity.) It was quite a surprise to me that the mussels found in dinosaur layers look similar to modern forms.

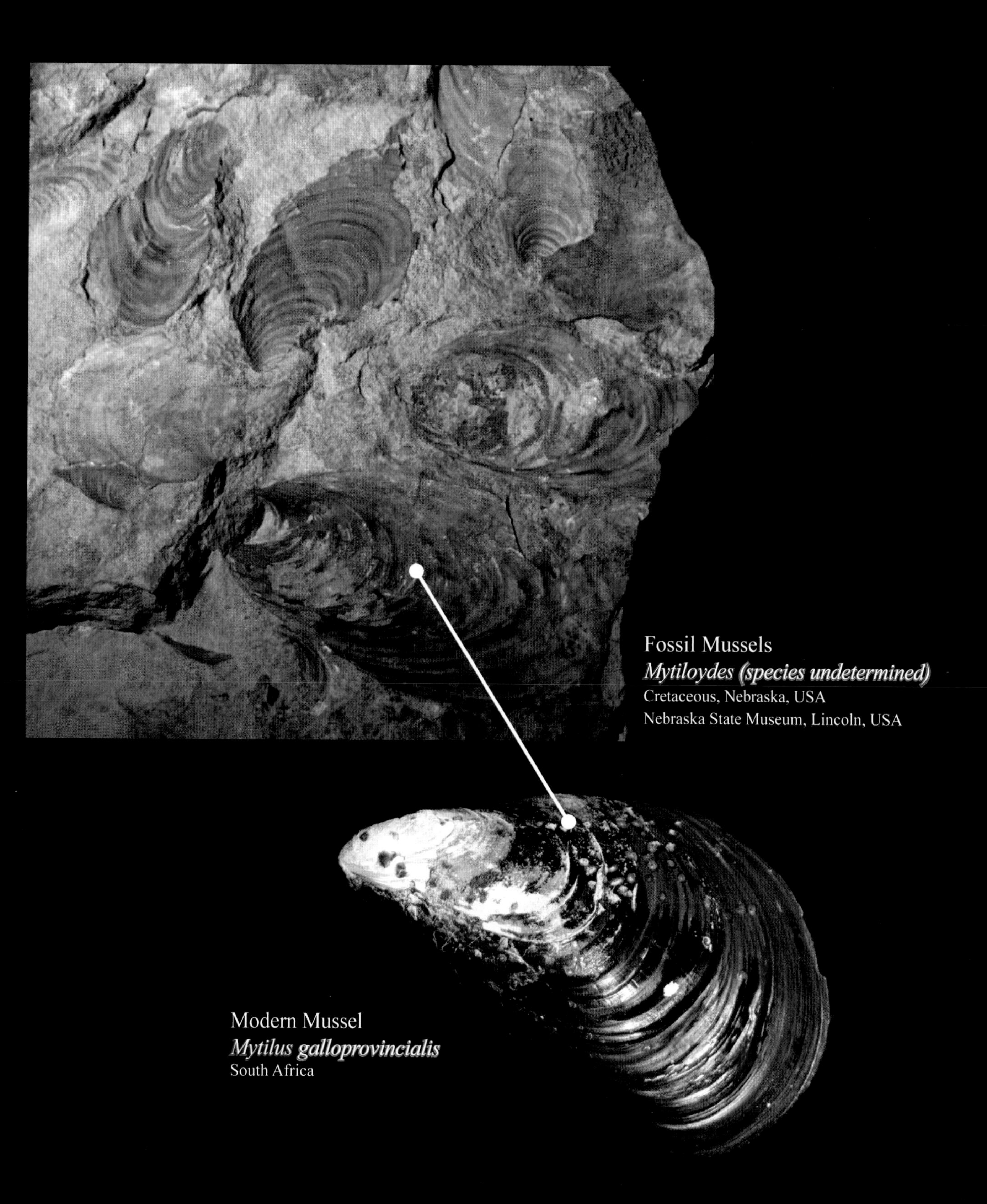

Fossil Mussels
Mytiloydes (species undetermined)
Cretaceous, Nebraska, USA
Nebraska State Museum, Lincoln, USA

Modern Mussel
Mytilus galloprovincialis
South Africa

Cockscomb Oyster

Compare the fossil oyster below to a modern Cockscomb oyster. Other than color, which is not preserved in the fossil, and the photographic angle, I see no significant differences.

*See Appendix D: Color

Dinosaur-Era Cockscomb Oyster
Lopha (species undetermined)
Cretaceous, Charente, France
Carnegie Museum of Natural History, USA

Living Cockscomb Oyster
Lopha cristagalli

Summary

It appears that the most common and recognizable bivalve shellfish — scallops, clams, mussels, and oysters — have not changed dramatically since dinosaurs roamed the earth. This is a fulfillment of my prediction, that if evolution was not true, I would find modern animals in dinosaur rock layers.

Chapter 8

Snails

This chapter deals with the second major class of shellfish living today, snails (Phylum Mollusca) — shellfish with a single-coiled shell.

"At the most famous dinosaur site in the world, Dinosaur National Monument, I was surprised to see snails that looked like the ones in my aquarium."

Visitors looking at dinosaur bones in rock wall at Dinosaur National Monument.

Building at Dinosaur National Monument which covers dinosaur bones in rock wall.

Park Ranger touching dinosaur leg bone in rock wall.

Dinosaur National Monument, Utah

Side of building covering bluff.

Distant view of building at Dinosaur National Monument.

Freshwater Snails Found at Dinosaur National Monument

These four fossil snails below were found at Dinosaur National Monument along with a *Stego-saurus* dinosaur and a host of long-necked sauropod dinosaurs. Compare these fossilized snails to a living freshwater snail at the bottom of this page. To me, the structure of the snail has not changed to any great degree. What is your impression?

Dinosaur-Era Snails
Dinosaur National Monument
Vernal, Utah, USA

Living Snail
Author's Home Aquarium

Saltwater Snails

Compare this fossil snail (below, left), found at Petrified Forest National Park in Arizona along with the dinosaur *Coelophysis*, to a modern saltwater snail (below, right).

Petrified Forest National Park

Coelophysis Dinosaur

Dinosaur-Era Snail
Top broken off
***Lioplacodes* (species undetermined)**
Triassic, Petrified Forest National Park, USA
Arizona, USA[1]

Modern Saltwater Snail
Babylonia perforata
East China Sea

Moon Snails

This fossil snail (below, right) was found in Cretaceous rock layers in Australia. Compare it to a living moon snail found nearby on an Australian beach (bottom, left). Do you think the differences between these two snails warrant a different genus and species name?

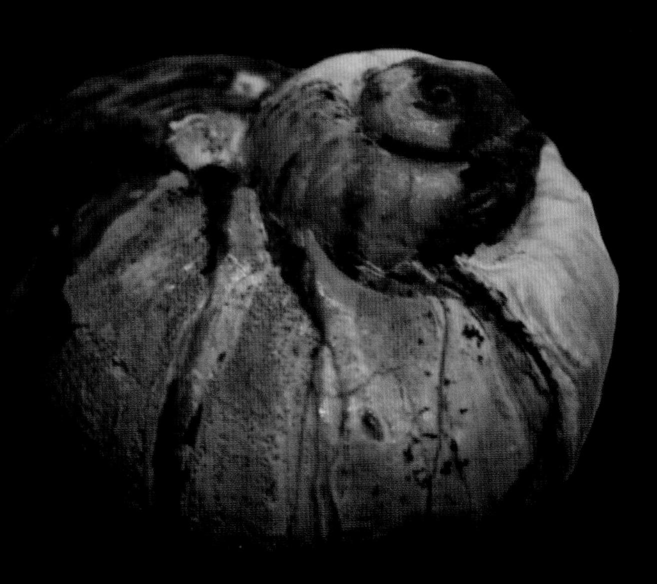

Dinosaur-Era Snail
Euspira reflecta
Cretaceous, South Australia
South Australian Museum, Adelaide

Modern Moon Snail
Polinices conicus
South Australia Beach

See Appendix A: *The Influence of Fossil Orientation*

Modern slit shells have a "slit" in the last ring of the shell, thus their name. The slit in the most delicate part of the shell makes it very vulnerable to breaking.

Modern Slit Shell
Mikadotrochus hirasei
Found off coast of Taiwan

Dinosaur-Era Shells
Conotomaria minacis
Cretaceous, Australia
Victoria Museum, Melbourne, Australia

Summary

Many snails have been found in dinosaur layers that look like modern freshwater and saltwater varieties.

Other Types of Shellfish

This chapter addresses the other smaller classes of the Phylum Mollusca living today, such as the chambered nautilus, tusk shells, and sea cradles. It also addresses lamp shells, which are members of the more obscure Phylum Brachiopoda.

Chapter 9

"After seeing the museum display, I concluded that the modern nautilus did not live during dinosaur times. Now I am not so sure."

Chambered Nautilus and Dinosaurs

As children, my brothers, sister, and I were captivated by the variety and colors of shells we found on the beaches of Florida while on family vacations. On one occasion, my brother bought a beautiful orange and white nautilus shell at a local shop (above, right), which laid on a shelf in our bedroom for years, eventually making a visual imprint in my mind.

You can imagine how surprised I was to come upon a fossil (below, right) at the California Academy of Sciences Natural History Museum, which looked similar to the one in our bedroom but was identified as extinct. I assumed the fossil must have had some unusual internal anatomy that I simply could not see. I gave it a nod and concluded that the modern nautilus shellfish did not live during dinosaur times.

Modern Nautilus
Nautilus pompilius
Collected in the Philippines

Dinosaur-Era Nautiloid
Cymatoceras texanum
Cretaceous,Texas, USA
California Academy of Sciences, USA

Eight years later, I came across *another* fossil "nautilus-like" shellfish from the dinosaur era, but this one was cut in half, allowing me to take a look inside and make a judgment for myself.

I placed a modern nautilus shell (cut in half) next to the fossil. I could now see the inside of both.

Note that the tube-like siphuncle (marked with an "X") is well-preserved in the fossil and cuts through the center of each chamber. In the modern nautilus shell, the soft-tissued siphuncle and other fleshy parts have been removed but can be inferred by the indentations in the wall of each chamber.

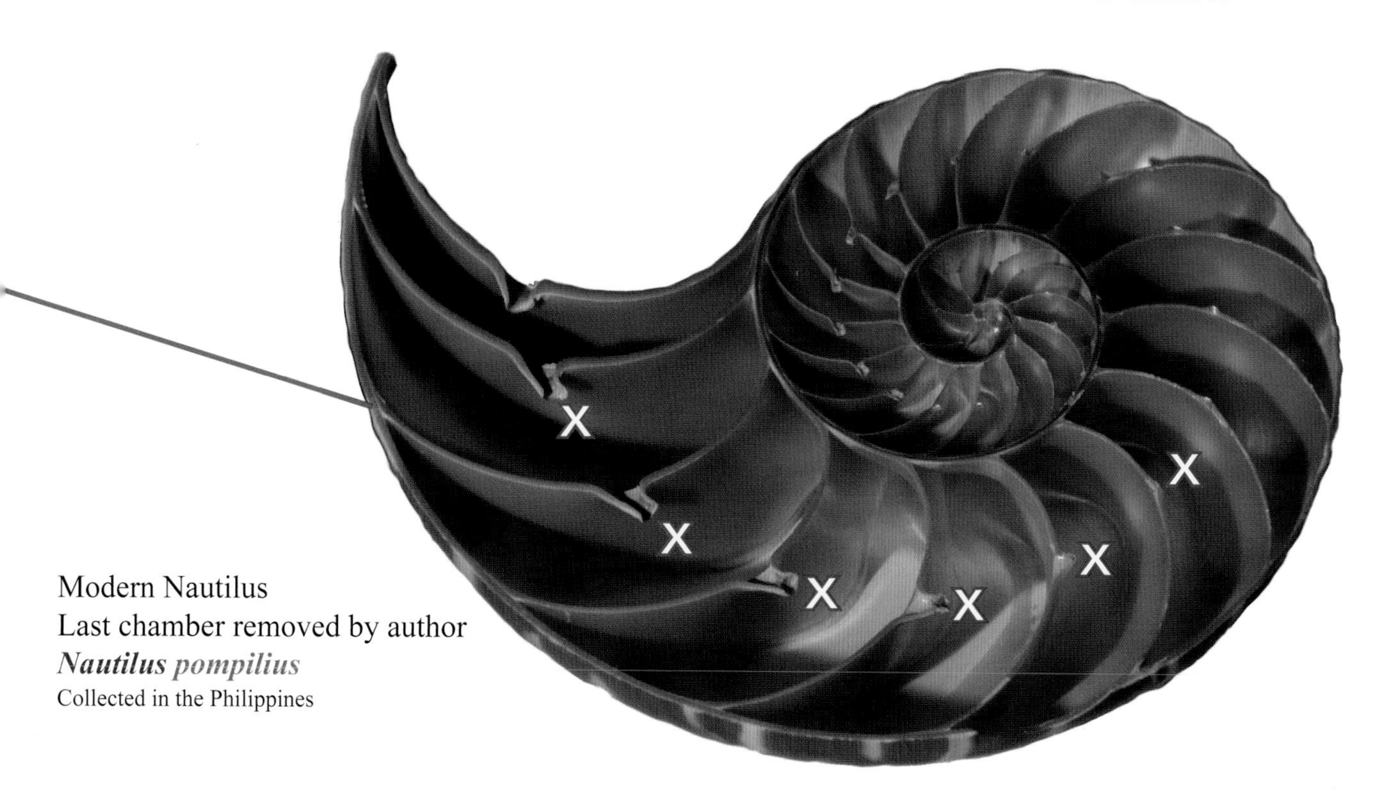

Modern Nautilus
Last chamber removed by author
Nautilus pompilius
Collected in the Philippines

X = siphuncle

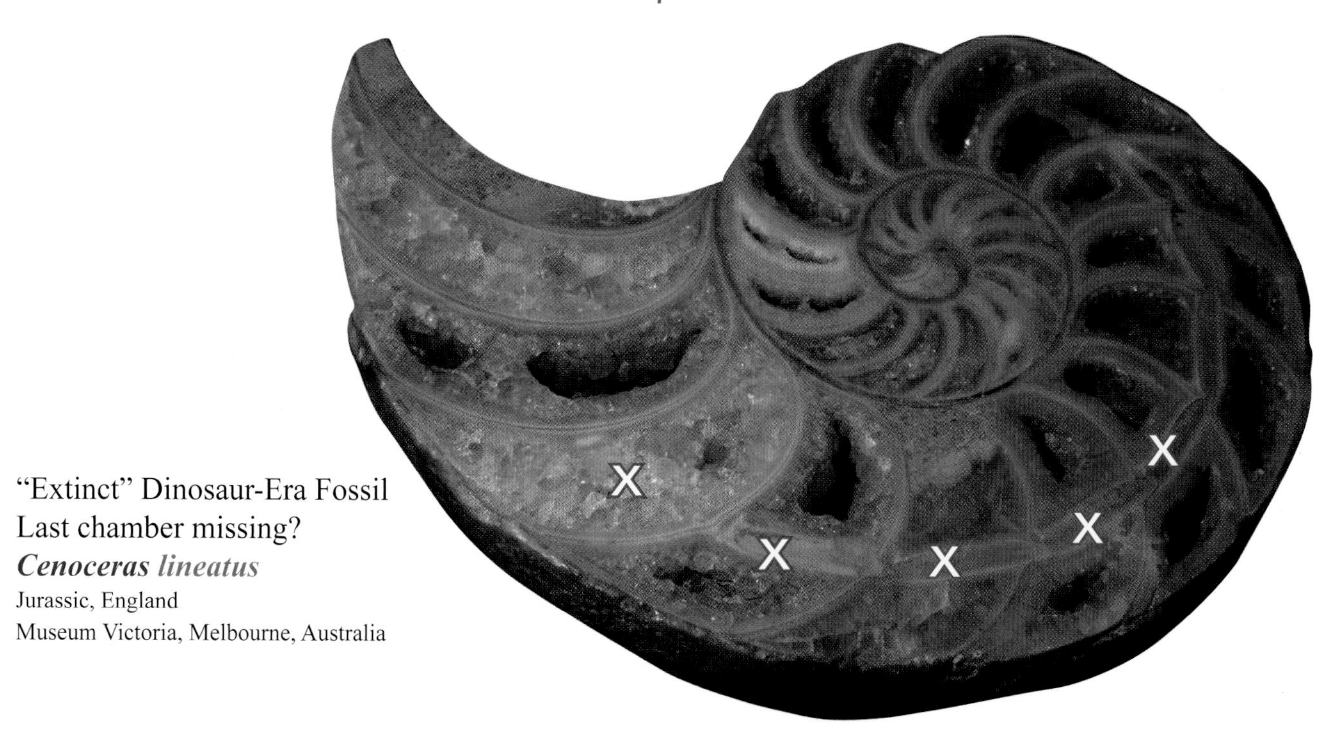

"Extinct" Dinosaur-Era Fossil
Last chamber missing?
Cenoceras lineatus
Jurassic, England
Museum Victoria, Melbourne, Australia

Dinosaur-Age Elephant Tusk Shells
Dentalium gracile
Cretaceous, South Dakota, USA
Carnegie Museum of Natural History, USA

Modern Elephant Tusk Shell
Dentalium (species unidentified)
Collected in the Philippines

Tusk Shells

As the name implies, tusk shells look like elephant tusks, but are only inches long. Look at these marine tusk shells found in dinosaur-bearing rock layers in South Dakota (top of page) and compare them to the modern variety (bottom of page).

Lamp Shells

Lamp shells look like clams but are placed in a completely different phylum (Brachiopoda) because of matters dealing with shell symmetry. Compare the top shell of this living lamp shell to a fossilized lamp shell below.

Modern Lampshell
Top and bottom shells
Rhynchonella psittacea
White Sea, Russia

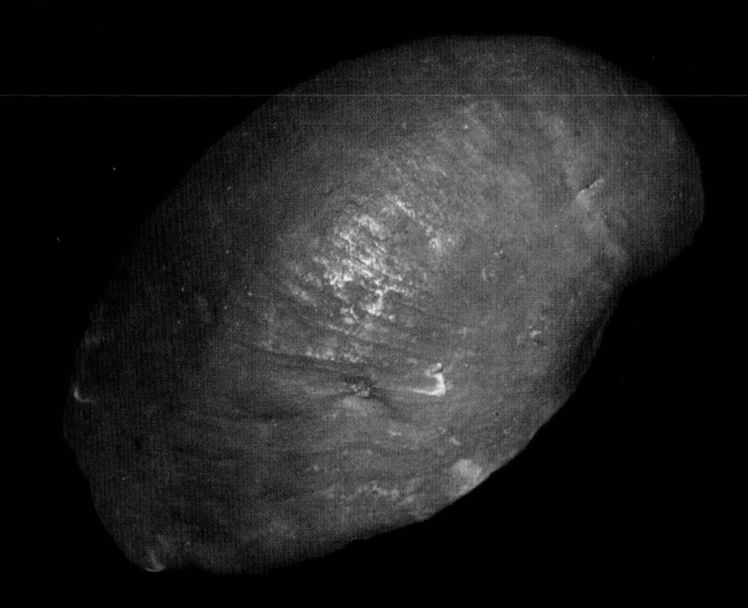

Dinosaur-Era Fossil Lampshell
Top shell only
Terebratula simplex
Mesozoic, England
Museum Victoria, Melbourne, Australia

Sea Cradles

Sea cradles are the fifth and last major class of shellfish living today. They are marine mollusks with eight interlocking plates. Sea cradles are living today and have been found in dinosaur rock layers.[1] These fossils are very rare and usually consist of a few fragments of the interlocking plates.

Modern Sea Cradle

Summary

Examples from *all* five major classes of shellfish living today in the Phylum Mollusca have also been found in dinosaur rock layers, including bivalves (Class *Bivalvia*), snails (Class *Gastropoda*), chambered shellfish (Class *Cephalopoda*), tusk shells (Class *Scaphopoda*), and sea cradles (Class *Polyplacophora*).[2,3] Besides these, even lamp shells of the more obscure Phylum Brachiopoda have been found in dinosaur rock layers.

Modern Sea Cradle
Stenosemus albus
Russia

Chapter 10

Worms

This chapter deals with segmented worms (Phylum Annelida).

"I was baiting my hook, and it occurred to me that the worm in my hand looked like the fossil I saw in Australia."

Earthworms

The fossil worm pictured on the next page, labeled an Oligochaete, was found in a dinosaur fossil layer in Australia. Months after seeing this fossil at the Museum Victoria in Melbourne, Debbie and I went fishing at Table Rock Lake in Missouri. As I was sitting in the boat, baiting my hook, it occurred to me that the night crawler in my hand was similar to the fossil worm. (As you can see, the challenge is never far from my mind.)

Since we go very few places without a digital camera, I put another night crawler in my hand and took a picture of it for later analysis at my home. I then turned my attention to the beautiful bass I caught.

Fossil Worm Found in Australia

Dinosaur-Era Worm
Oligochaete
Cretaceous, Australia
Museum Victoria,
Melbourne, Australia

Modern Earth Worm
Oligochaete
Missouri, USA

Tube Worms

Tube worms live today in the oceans and are quite beautiful. My favorite is this "Christmas Tree" tube worm (top right) which I photographed in the U.S. Virgin Islands off of St. John. The Christmas Trees are the only visible part of the tube worm. The rest of the worm, the casing, is buried in the coral (right). If you wave your hand over these colorful "trees," in the blink of an eye they quickly retract back into their casing.

While filming at the Nebraska State Museum in Lincoln, Debbie photographed this fossil tube worm (bottom, right) found in a dinosaur layer. Compare the fossil tube worm to the modern tube worm.

Christmas Tree Tube Worm
Spirobranchus
St. John, U.S. Virgin Islands

Modern Tube Worm Casing
Unidentified species

Dinosaur-Era Tube Worm
Unidentified species
Mesozoic, Unidentified Location
University of Nebraska State Museum

Summary

In the animal phylum of segmented worms (Annelida), modern-appearing examples of both classes have been found as fossils in dinosaur rock layers, including freshwater and saltwater varieties.[1]

Chapter 11

Sponges and Corals

This chapter deals with sponges (Phylum Porifera) and corals (known as Phylum Cnidaria by many zoologists, and Phyla Coelenterata to paleontologists).

"The sponges I see in the Caribbean look like the fossils."

When snorkeling in the coral reefs of the Caribbean, my favorite animals to photograph are the sponges because they are so abundant, so colorful, and they always "hold still" in order to set up a shot. I photographed the sponge on the previous page in St. Thomas.

To me, the two sponges on this page, one living and one from the dinosaur rock layers, look very similar.

Dinosaur-Era Fossil Sponge
Purisphonia clarkei
Cretaceous, South Australia
South Australian Museum, Adelaide

Demosponges

Living Demosponge
Haliclona aquaeductus
Russia

Glass Sponge Found in the White Cliffs of Dover

The famous White Cliffs of Dover in England have borne many spectacular fossils, including glass sponges. These layers were deposited at the same time that the dinosaurs were alive. The glass sponge below was found here.

Modern Glass Sponge
***Hyalonema* species undetermined**
Florida, USA

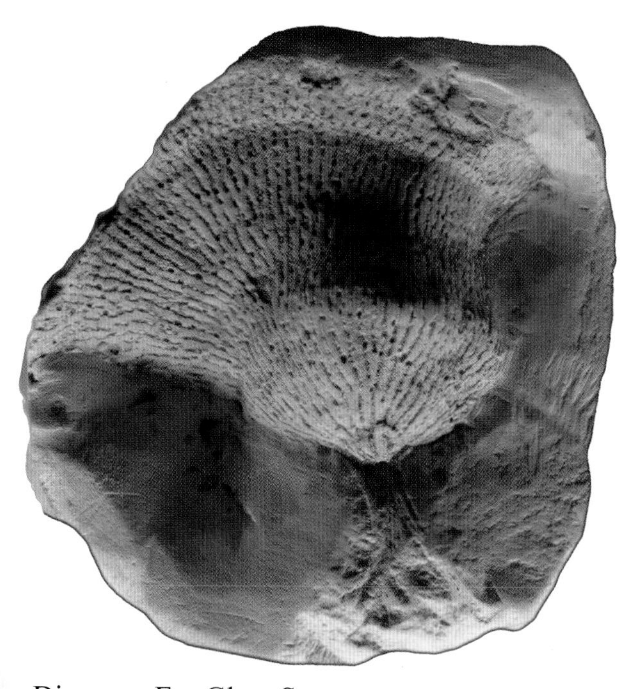

Dinosaur-Era Glass Sponge
Ventriculies radiatus
Cretaceous, England
Carnegie Museum of Natural History, USA

Humpback Coral

Next look at this fossil humpback coral found in Austria and compare it to the living humpback coral. Although they were photographed at slightly different angles, to me, they look similar.

Living Humpback Coral
Cycloseris vaughani
Hawaii, USA

Dinosaur-Era Humpback Coral
Cyclolites undulata
Cretaceous, Austria
Carnegie Museum of Natural History, USA

Other Corals

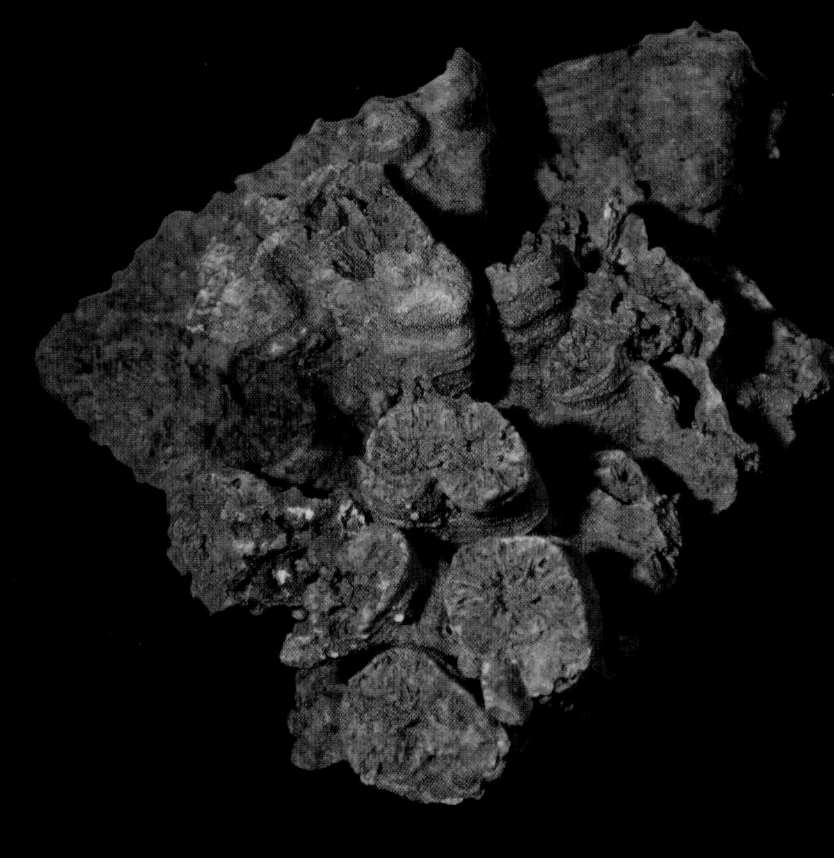

Dinosaur-Era Coral
Unidentified species
Mesozoic, Germany
Museum Victoria, Australia

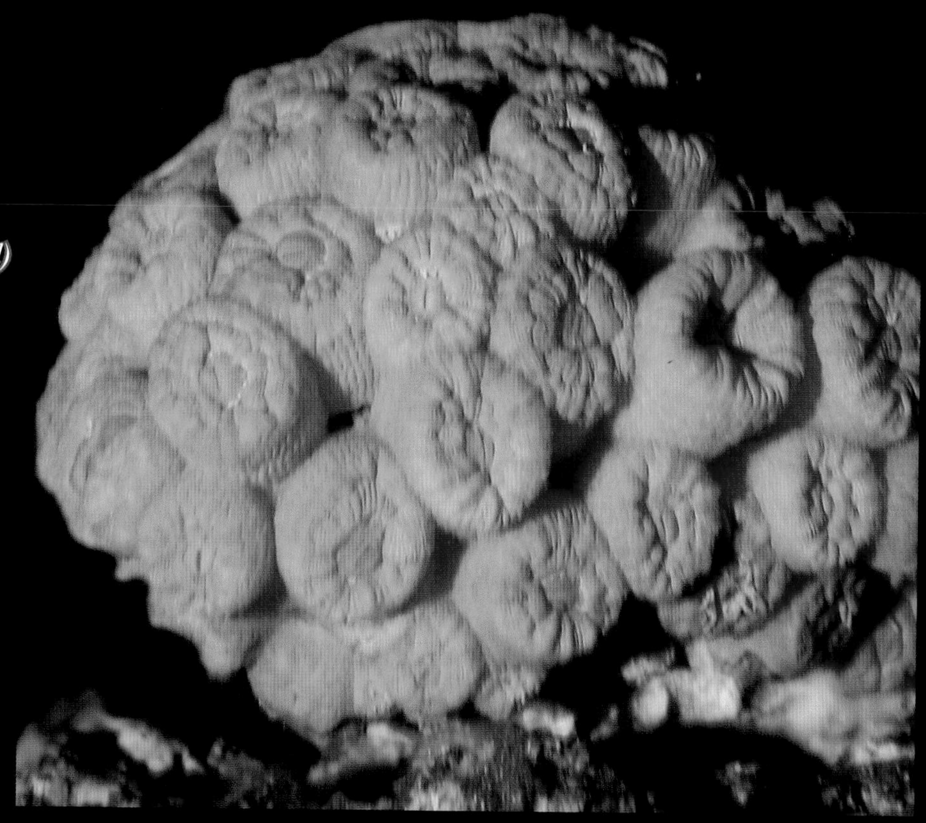

Living Orange Cup Coral
Tubastrea (species unidentified)
Australia

Summary

Finding fossils of all three classes of sponges living today,[1,2] and a variety of corals, including hard and soft corals,[3] was similar to my experiences with echinoderms, arthropods, and shellfish. Now I wanted to see if I could find the last major phylum of animals — the vertebrates. Would I be able to find modern-appearing fish, amphibians, reptiles, birds, and mammals at the dinosaur dig sites? For me, this would be the real test.

Bony Fish

This chapter addresses bony fish, the largest class of fish living today (Phylum Chordata).

Chapter 12

"Imagine my surprise when I visited dinosaur dig sites and found modern-appearing fish."

Introduction

During the reading phase of my research from 1979 to 1997, I developed the impression that ancient fish were, in general, different from the fish living today. But when I actually visited the dinosaur dig sites, I found many modern-appearing fish with strange names assigned to them. The first instance of this occurred at Dinosaur Provincial Park in Alberta, Canada, on our very first trip. Here on display at the Field Station were the fossil scute and skull bones of a sturgeon.

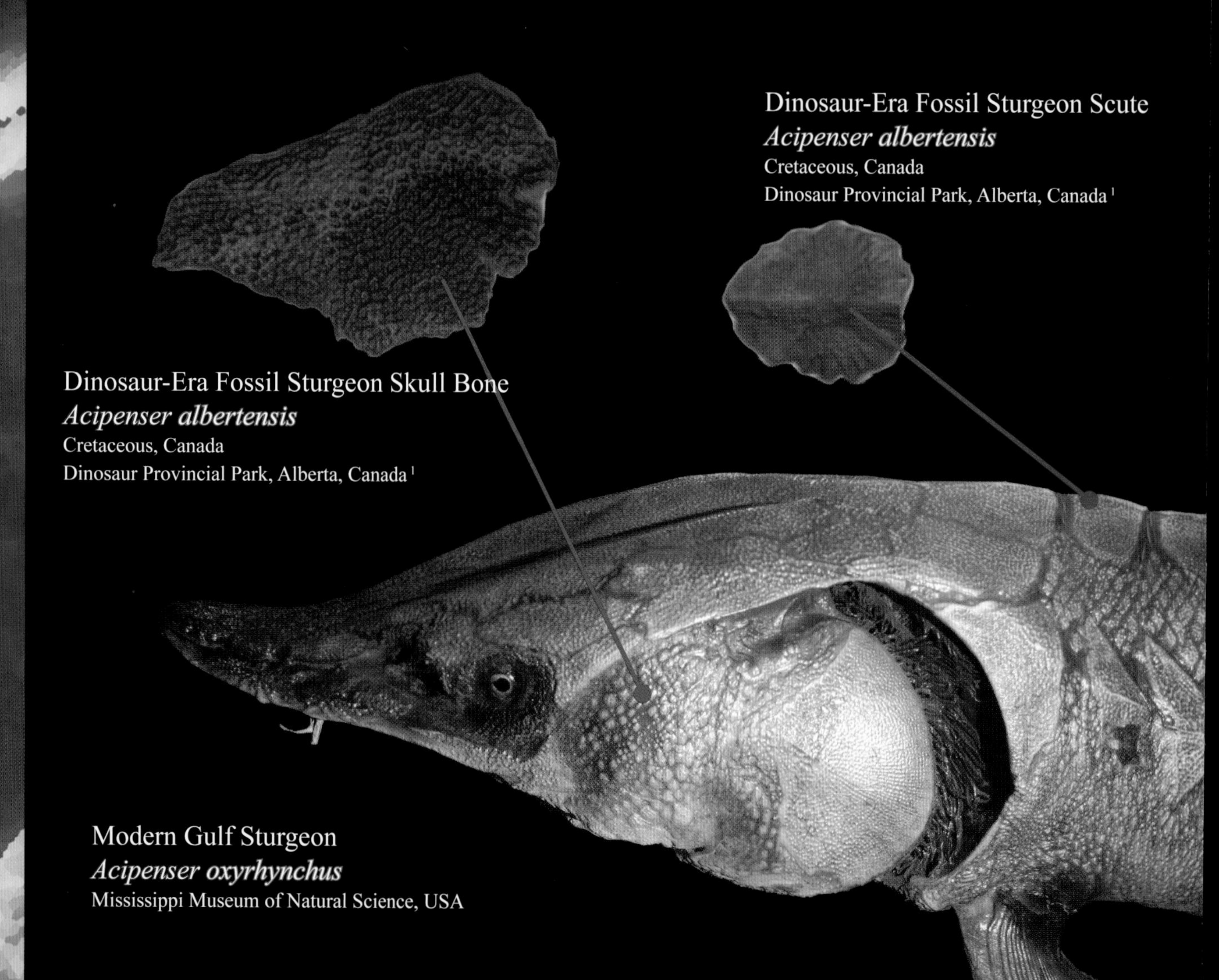

Dinosaur-Era Fossil Sturgeon Scute
Acipenser albertensis
Cretaceous, Canada
Dinosaur Provincial Park, Alberta, Canada [1]

Dinosaur-Era Fossil Sturgeon Skull Bone
Acipenser albertensis
Cretaceous, Canada
Dinosaur Provincial Park, Alberta, Canada [1]

Modern Gulf Sturgeon
Acipenser oxyrhynchus
Mississippi Museum of Natural Science, USA

Sturgeon Found at Dinosaur Provincial Park

Dinosaur Provincial Park
Alberta, Canada

Museum Field Station inside
Dinosaur Provincial Park

Coelacanth

Most of the experts we interviewed supported the idea that many dinosaur-era fish appeared relatively unchanged but insisted they were different species. Looking carefully at the two fish on the next page — noting the shape of the fish, the fins, and the tail — I find it hard to believe that (1) they are from different genus groups, and (2) they could not reproduce. They appear to me more similar than a Beagle and a Pug, which are the same species.

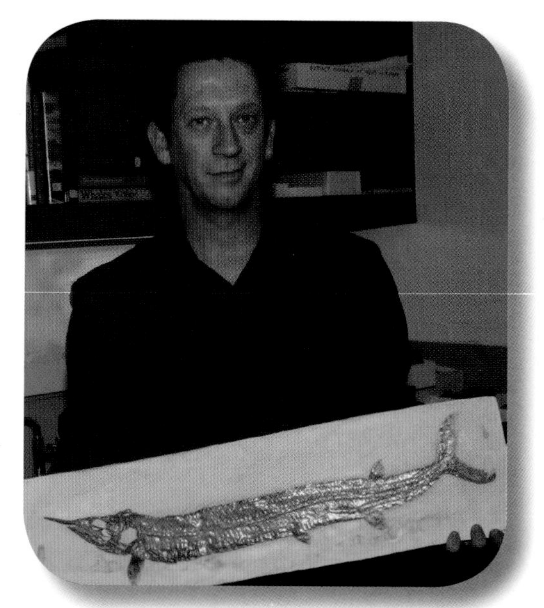

"Coelacanths...quickly reached their modern form by the start of the Mesozoic [dinosaur] *Era, and have remained relatively unchanged ever since."* [2]

— Dr. Long

Dr. John Long is a proponent of evolution and author of ***The Rise of Fishes: 500 Million Years of Evolution.***

Coelacanths Are Living Today and Lived with Dinosaurs, Pterosaurs, and Mosasaurs.

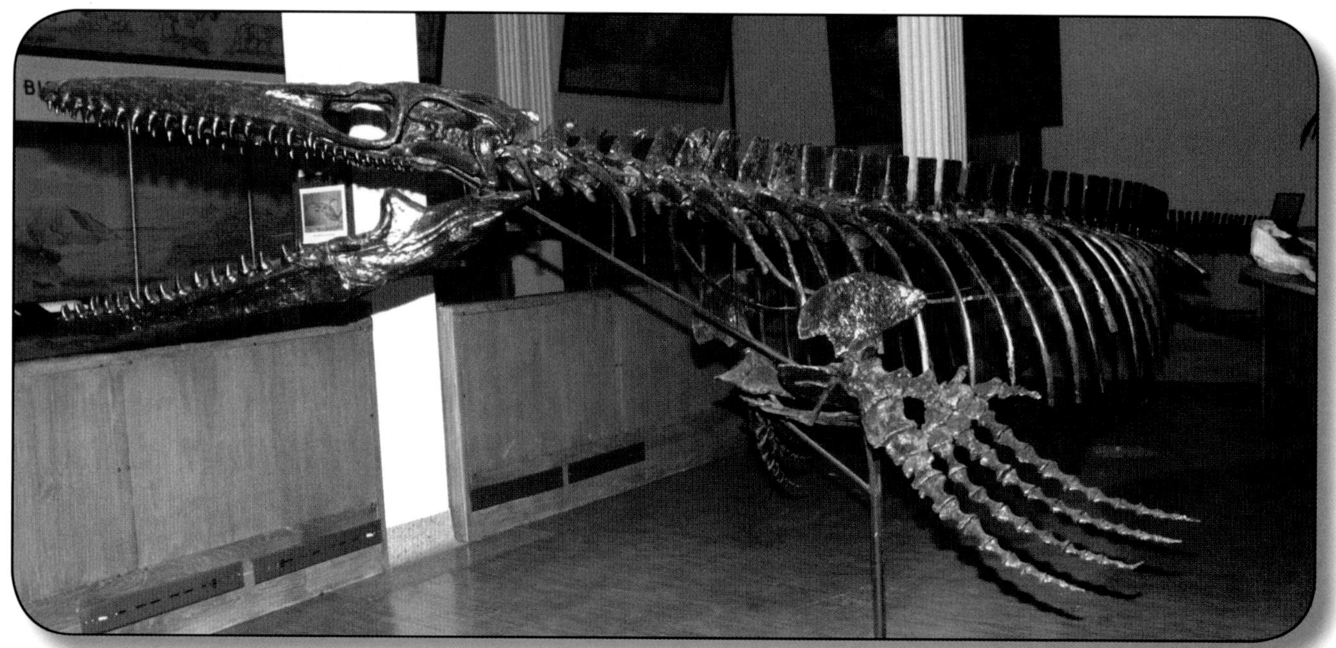

Mosasaur
Mosasaurus *condon*
Cretaceous, South Dakota, USA
South Dakota School of Mines and Technology, USA

Modern Coelacanth
Latimeria chalumnae
Caught in Indian Ocean, Comoro Islands
California Academy of Sciences, USA

Dinosaur-Era Coelacanth
Coelacanthus pencillatus
Jurassic, Eichstatt, Germany
Harvard Museum of Paleontology, USA

Salmon

Dr. David Weishampel (Johns Hopkins University), an expert in dinosaur evolution, suggested that salmon-like fish were found in European dinosaur layers where he was conducting a dinosaur dig. [3] Unfortunately, he did not have the salmon-like fossils for us to photograph, but other scientists have also reported finding salmon and other modern fish types at their dinosaur dig sites.

"With the dinosaurs that we find in Transylvania, we find other kinds of vertebrate fossils as well. We find fishes, bony fishes, say like ***salmon,*** *or everybody's general sense of what a fish looks like. They're called 'bony' fishes."* [3]

— Dr. Weishampel

Dr. David Weishampel, anatomist and paleontologist, Johns Hopkins University, and lead editor of the encyclopedic reference book ***The Dinosauria****.*

"Many of the bony fishes, things like the salmon, the salmoniforms, have gone through [the event which killed off the dinosaurs] *unaffected, and today they give us a bounty of some of the best food we can eat."* [4]

— Dr. Long

Dr. John Long, Paleontologist and Head of Science, Museum Victoria, Melbourne, Australia.

Lungfish

*"Fossils of the Australian lungfish indicate that this species has **remained unchanged** in Australia for at least 100 million years...."* [5]

— Dr. Long

Dr. John Long, paleontologist and head of science, Museum Victoria, Melbourne, Australia.

Living Lungfish
Neoceratodus forsteri
Sydney Aquarium, Australia

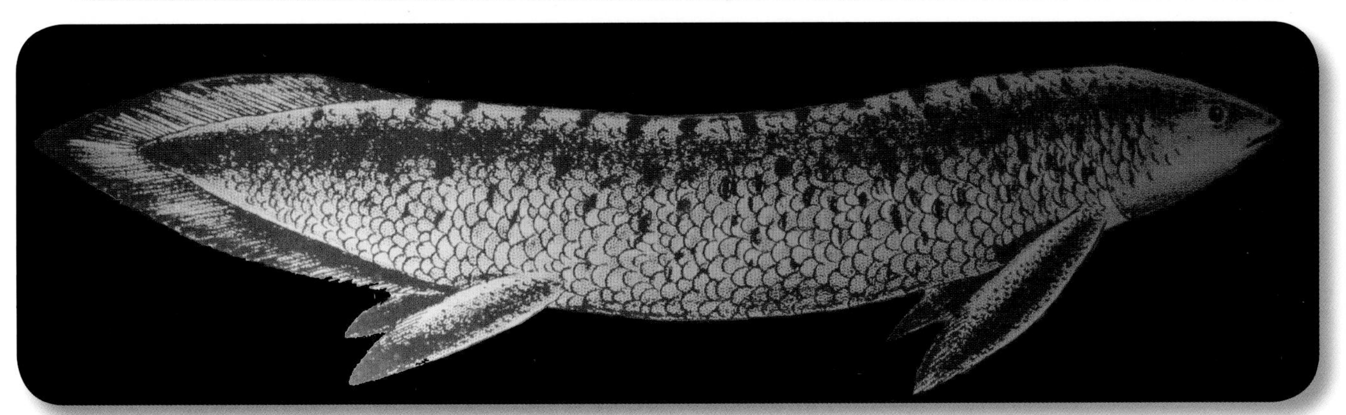

***See Appendix B:** *Use of Fossils Found below Dinosaurs*

Lungfish Found below Dinosaur Layer
Sagenodus periprion
*Permian, unidentified location
Carnegie Museum of Natural History, USA

Gar

I find it contradictory when scientists claim that fossil gars are a different species than the modern form even though the bones and scales of the fossil gar are "difficult to distinguish" from modern forms.

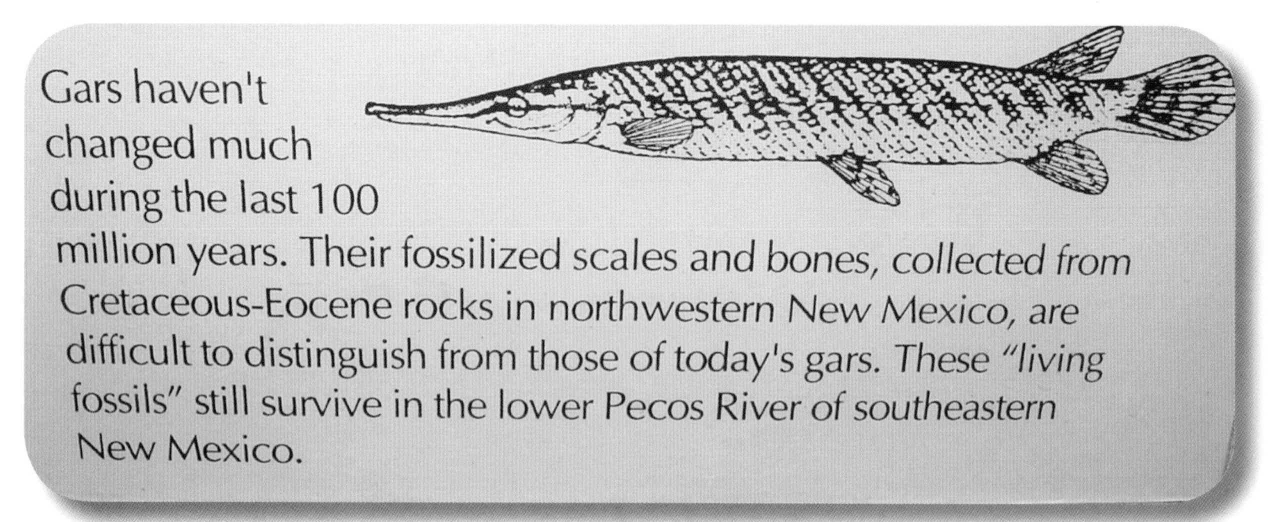

Above: *Museum sign at the New Mexico Museum of Natural History and Science in Albuquerque, New Mexico.*

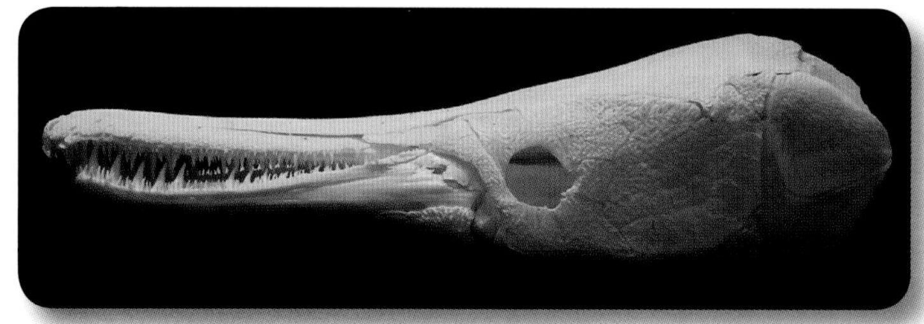

Modern Gar Skull
Lepisosteus** **(species unidentified)
Sam Noble Oklahoma Museum of Science and Natural History, USA

Modern Gar
Lepisosteus spatulata
New Mexico Museum of Natural History and Science, USA

Bowfin

At the bottom of this page is a reconstructed model of a bowfin fish. The Milwaukee Public Museum created this fish replica based on actual fossil bones found at a dinosaur dig site in Montana. The color of the fish and the size of the tail spot would not have been preserved in the fossil.* The artist's interpretation of these features are speculative and should be ignored. Focusing instead on the fins and body shapes of these two fish, one living and one from the dinosaur era, do you see change significant enough to warrant different genus and species names?

Living Bowfin
Amia calva
Milwaukee County Zoo, USA

Dinosaur-Era Bowfin Model
Kindleia fragosa
Cretaceous, Montana, USA
Milwaukee Public Museum, USA[6]

***See Appendix D:** *Assumptions in Fossil Reconstructions — Color*

Paddlefish

Modern Paddlefish
Polyodon spathula
World Aquarium, Missouri, USA

Interviews from
Evolution**: *The Grand Experiment***
video series

Dr. Clemens, from the University of California,

indicated that paddlefish also lived with dinosaurs.

Dr. Clemens: *"Sturgeon and paddlefish and alligator gar are forms that have been found in association with dinosaurs."* [7]

Dr. William Clemens, curator of the Museum of Paleontology, UC Berkeley

Torvosaurus
Jurassic, USA
Mesalands Community College
Tucumcari, New Mexico, USA

Eel

Living Eel
Unidentified genus
Vietnam

Dinosaur-Era Eel
Unidentified genus
Cretaceous, Lebanon
Chicago Field Museum, USA

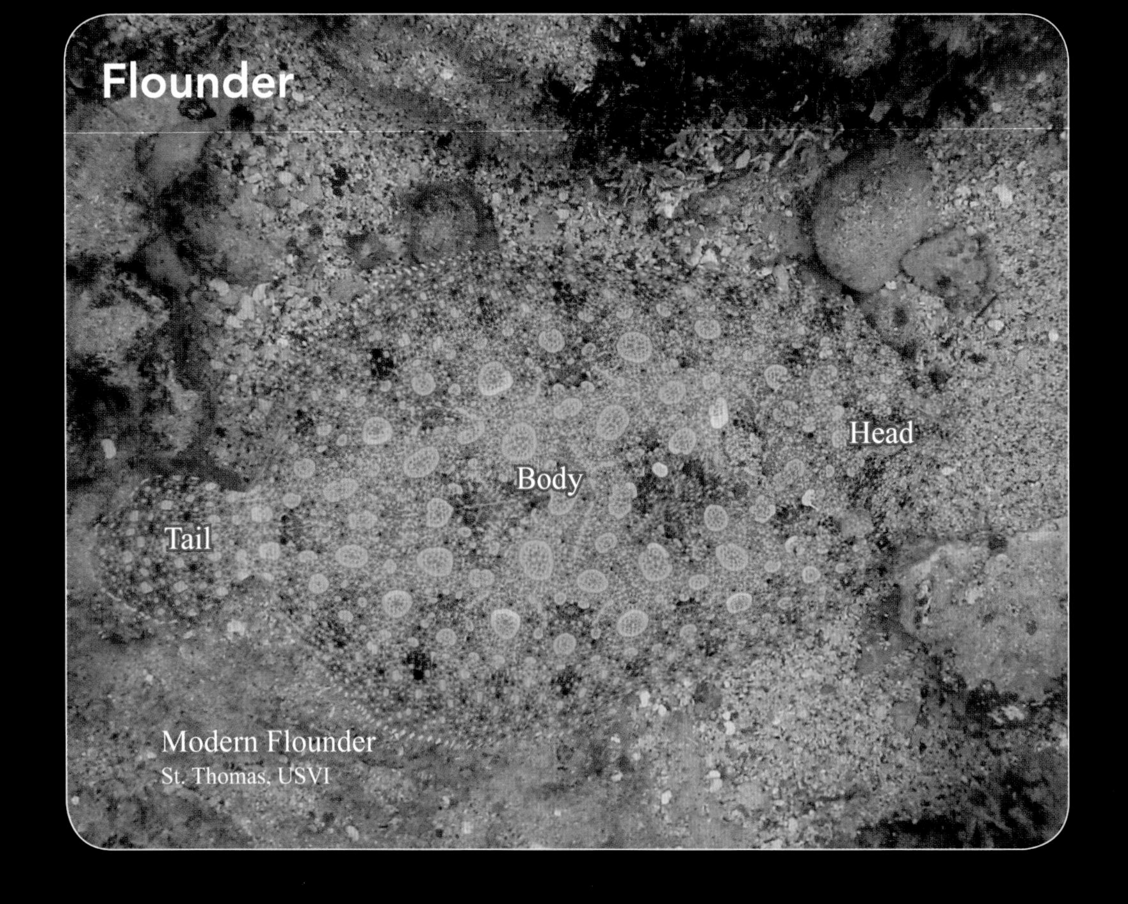

Modern Flounder
St. Thomas, USVI

I photographed this live flounder (above) in St. Thomas, U.S. Virgin Islands. If you look very carefully, you can see the outline of this spotted fish lying on its side. Its head is on the right and its tail is on the left.

This living fish would not be all that significant except for the fact that scientists at the New Mexico Museum of Natural History and Science in Albuquerque (right) reported finding flounder fish in dinosaur fossil layers, thus adding another example of a modern type of fish to my list. [8]

LIFE IN THE SHALLOW SEAS

Marine life in New Mexico's Cretaceous seas was similar to that in present-day oceans. Modern groups of sharks and rays appeared and flourished during the Cretaceous. Some fossil forms are almost identical to living species. Many bony fishes, such as flounder, also lived 75 million years ago.

3000-gallon saltwater aquarium
Constructed by
Fiber-Tech Engineering Inc. and
Art & Technology, Inc. 1988
Technical direction by
Robert Johnson

The following pages offer pictures of other bony fish. *I cannot be sure that any of these pairs are the same species and no one should necessarily conclude this.* In order to know that, I would need access to a series of skeletons from these living fish and access to these fossils using a microscope. That being said, I feel an obligation to share these fossils with you.

Scad

Dinosaur-Era Fish
Leptolepis dubia
Jurassic, Solnhofen, Germany
Carnegie Museum of Natural History, USA

Yellowstripe Scad
Selaroides leptolepis
Malaysia

Herring

Dinosaur-Era Herring
Clupea (species undetermined)
Cretaceous, Lebanon
Museum Victoria, Melbourne, Australia

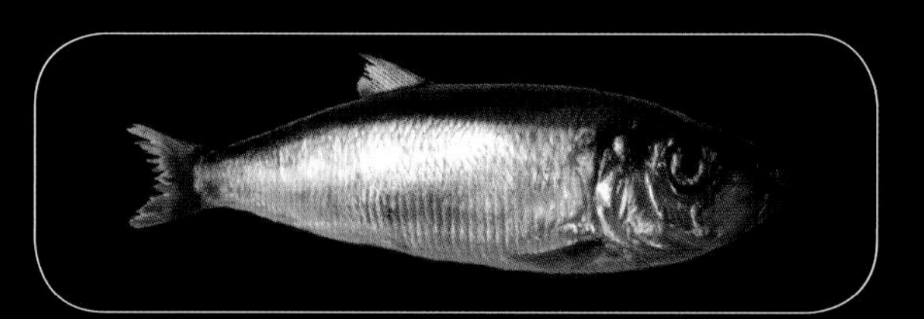

Atlantic Herring
Clupea harengus
Norway

Gissu

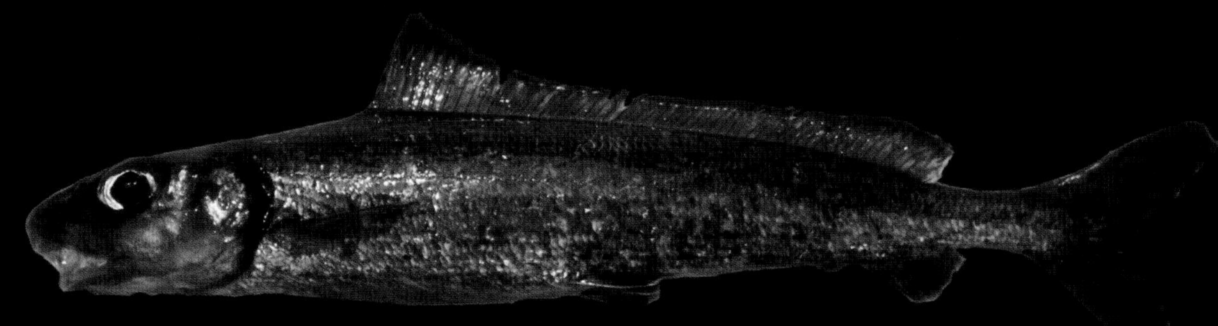

Living Gissu Fish
Pterothrissus gissu
Japan

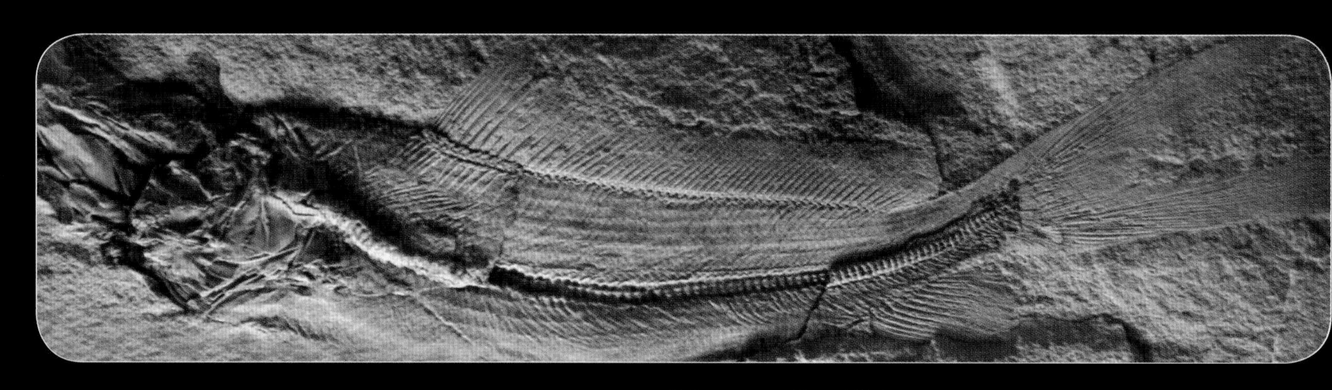

Dinosaur-Era Fish
Istieus gracilis

Orange Roughy

Dinosaur-Era Fish
Hoplopteryx lewisensis
Cretaceous, England
Carnegie Museum of Natural History, USA

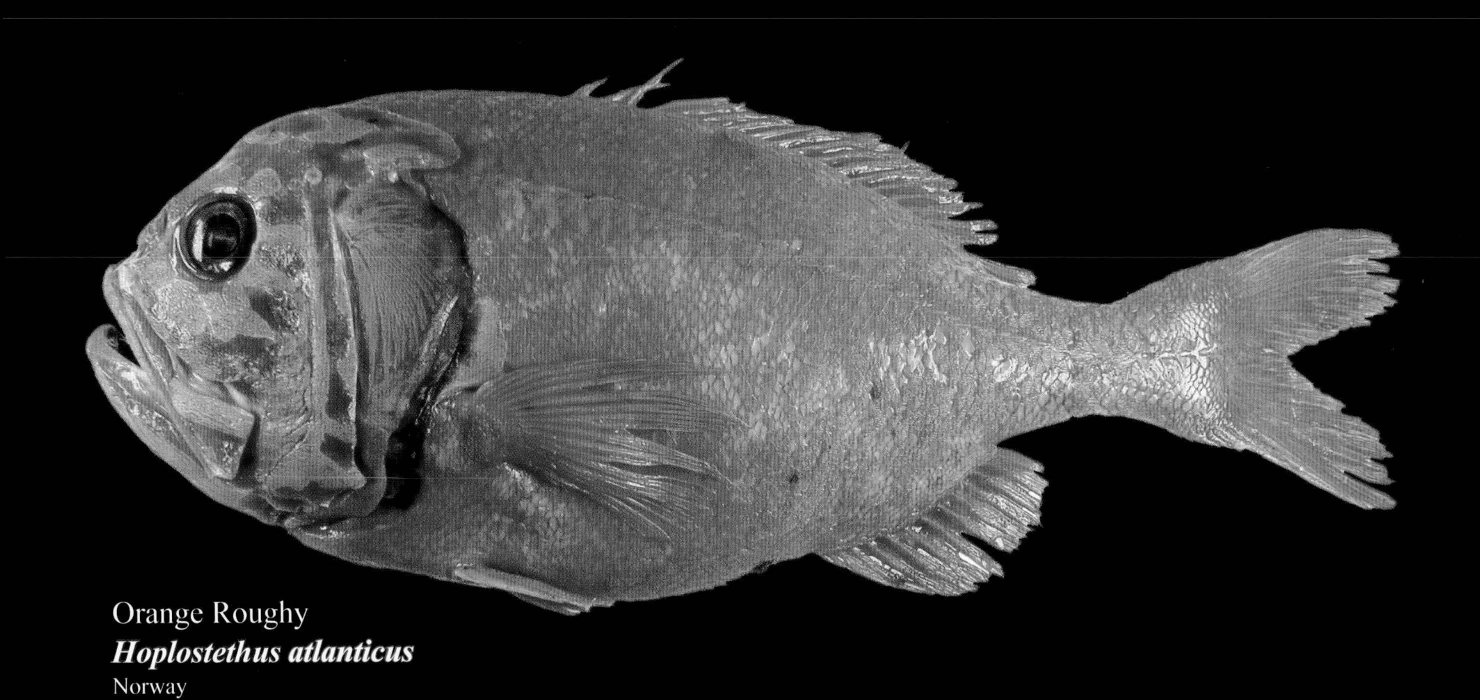

Orange Roughy
Hoplostethus atlanticus
Norway

Living Milkfish
Chanos chanos

Dinosaur-Era Fish
Aethalionopsis robustus
Cretaceous, Bernissart, Belgium
Royal Belgium Museum

Ladyfish

Ladyfish
Elops hawaiensis
Hawaii, USA

Dinosaur-Era Fish
Davichthys gardneri
Cretaceous, Lebanon
Chicago Field Museum, USA

Sardine

Sardine
Harengula jaguana
Florida Keys, USA

Dinosaur-Era Fish
Diplomystus brevissimus
Cretaceous, Syria
Carnegie Museum of Natural History, USA

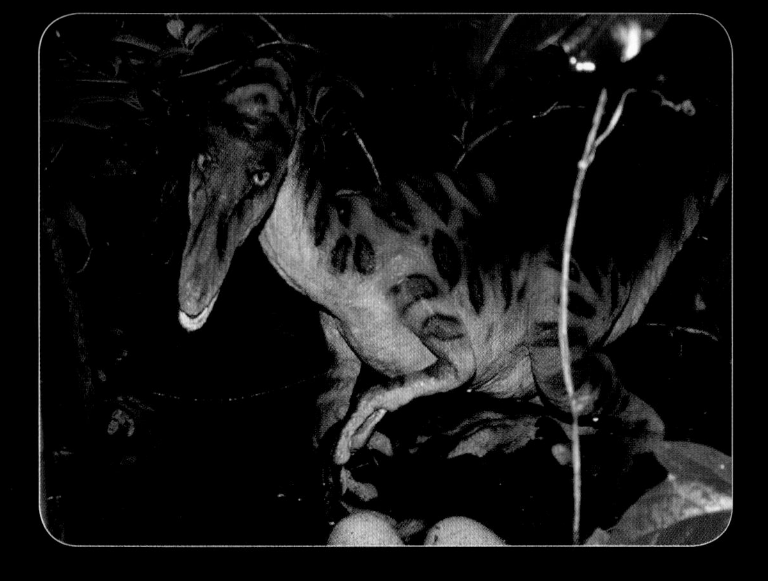

Left: *The dinosaur Troodon lived at the same time as many modern-appearing fish, including herring, gar, flounder, sturgeon, paddlefish, dogfish, salmon, and many, many others.*

Troodon
Troodon formosus
Cretaceous, Montana, USA
Missouri Botanical Gardens, USA
Living World Dinosaur Studio, Missouri, USA

Summary

My experiment predicted I would find modern bony fish in dinosaur fossil layers. I have found many examples of modern-appearing bony fish, including the ray-finned bony fishes (sturgeon, flounder, etc.) and the lobe-finned bony fishes (coelacanth). I am confident I could populate this list even further if I chose to.

Cartilaginous Fish

Cartilaginous fish (Phylum Chordata) have skeletons made of cartilage instead of bone and include sharks and rays.

Chapter 13

"The ancient sharks displayed at the Carnegie Museum were very modern in appearance."

Dinosaur Hall, Carnegie Museum of Natural History

Angel Shark

I felt like I had broken through a glass ceiling when I was able to interview and film at the Carnegie Museum in Pittsburgh. It was quite a thrill for me. Here on display were some of the best fossil specimens from around the world, including fossils from the dinosaur dig sites which I had already visited such as Solnhofen and Dinosaur National Monument in Utah.

Dinosaur-Era Shark
Squatina alifera
Jurassic, Eichstatt, Germany
Carnegie Museum of Natural History, USA

Even at this most prestigious museum, I found myself questioning evolution since I saw sharks and rays that appear similar to modern forms. Let me show you some examples.

Pacific Angel Shark
Squatina californica
Oregon, USA

Scientific illustrators at the Carnegie Museum drew this fossil shark (above) based on fossils found in Jurassic dinosaur layers in Germany. Compare the drawing to the living Pacific Angel Shark (left).

Dinosaur-Era Ray
Belemnobatis sismondae
Jurassic, Solnhofen, Germany
Carnegie Museum of Natural History, USA

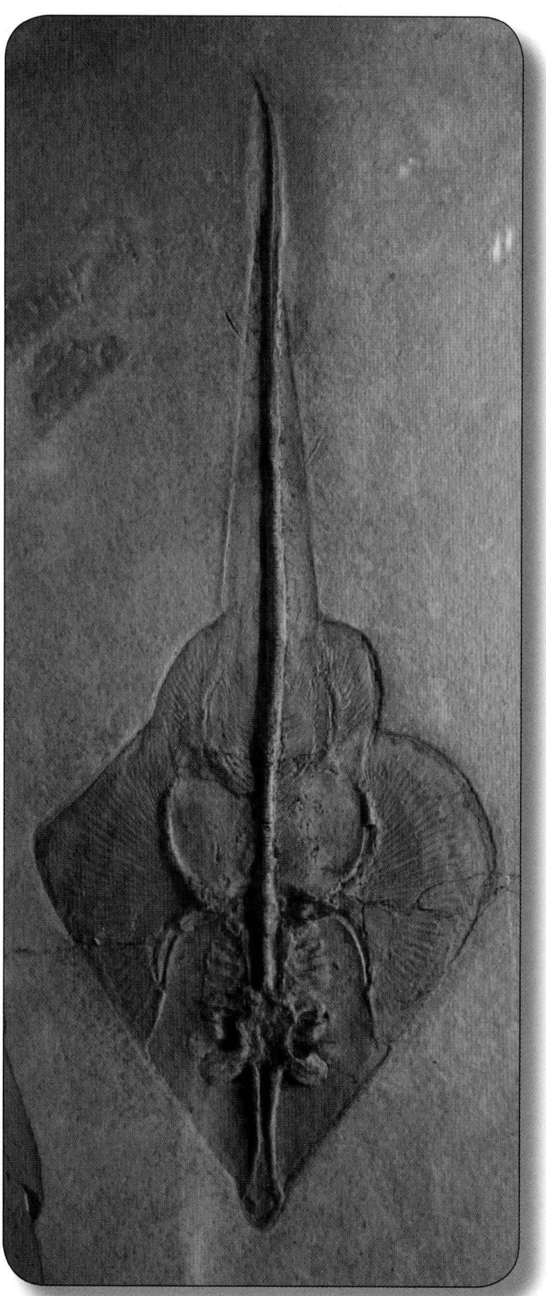

Modern Era Ray
Rhinobatos productus
Malibu, California, USA

Shovelnose Ray

Now look at this fossil ray (above, left). Compare this fossil to a living Shovelnose Ray caught off the coast of California. To me, these two fish look nearly the same.

Mystery Shark: Unusual Head, Tail, Fins, and Genus

Another cartilaginous fish I encountered (above) at the Carnegie Museum was labeled *Cestracion zitteli*; one I had never heard of before. I noted that this fossil fish was a member of the fish family Heterodontidae. A few years later, while filming in Australia, I took a photograph of a modern Port Jackson Shark (top of next page), which is also in the fish family Heterondontidae. When I placed the Port Jackson Shark photo next to the fossil photo, it did not match, not even close. All of that work was for nothing! But something kept bugging me about the fossil — it was the tail. I noted that the bones in the tail of the fossil (this page) pointed *down but* the bones in the tail of the modern Port Jackson Shark (top of next page) pointed *up*. I could also see that the fins did not line up, fin for fin. I almost dropped it, but it kept bugging me and I kept returning to it, over and over again, like a jigsaw puzzle. Then it hit me. *Maybe the museum displayed the fossil upside down.*

Using my computer, I flipped the fossil shark and, lo and behold, the tails now matched, as well as all of the fins and the fin spines. (See comparison on next page.)

Above: *Dinosaur Hall at the Carnegie Museum with a long-necked dinosaur in the foreground and dinosaur-era fossil sharks and fish along the outside wall.*

Port Jackson Shark and *Cestracion zitteli* (Flipped)

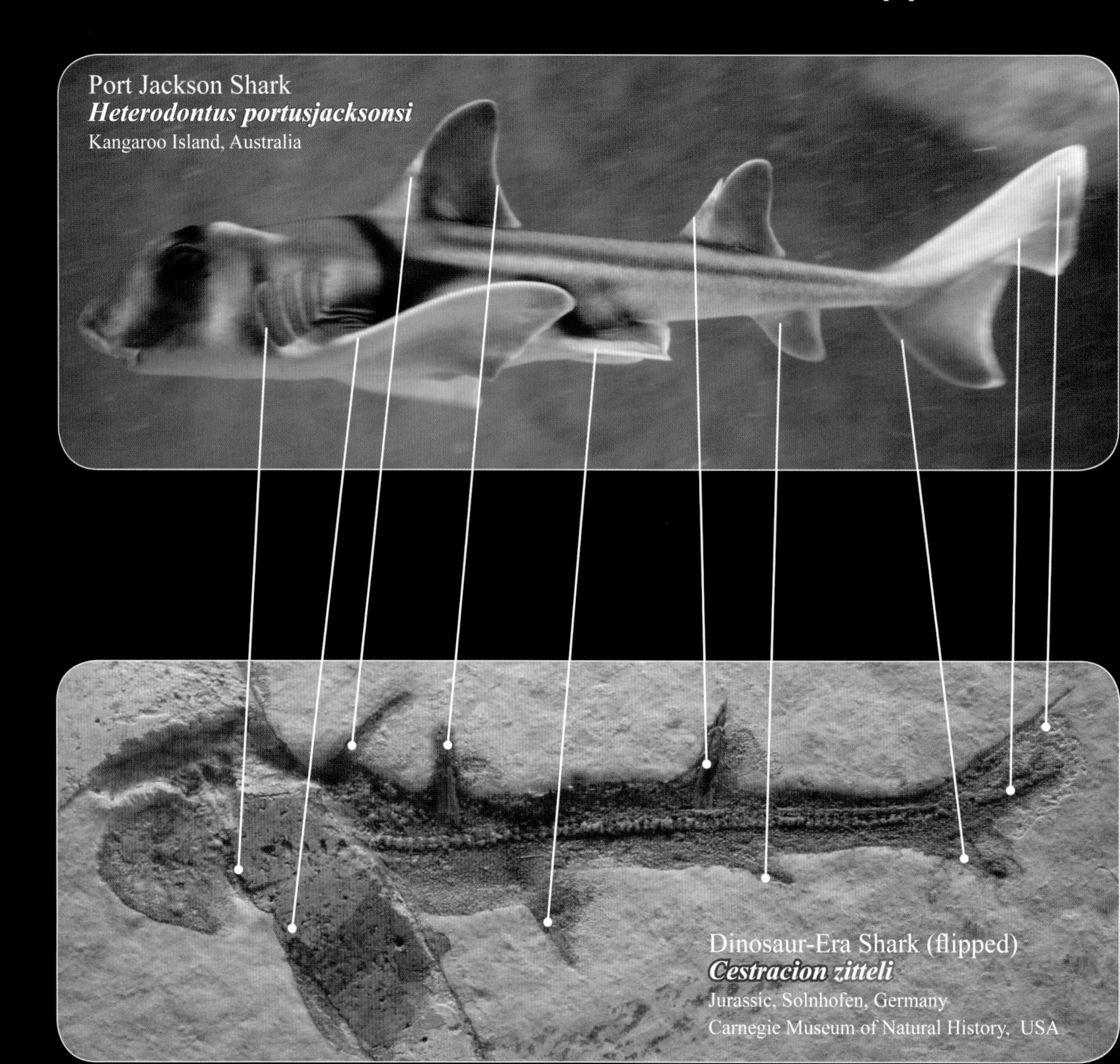

Goblin Shark

I photographed this "extinct" fossil shark tooth (bottom of page) at the Albuquerque Natural History Museum. After some research, I photographed a modern Goblin Shark tooth and put the photos side by side. Frankly, I cannot see any significant differences between the "extinct" shark tooth and modern Goblin Shark teeth. Now look at the genus names in blue!

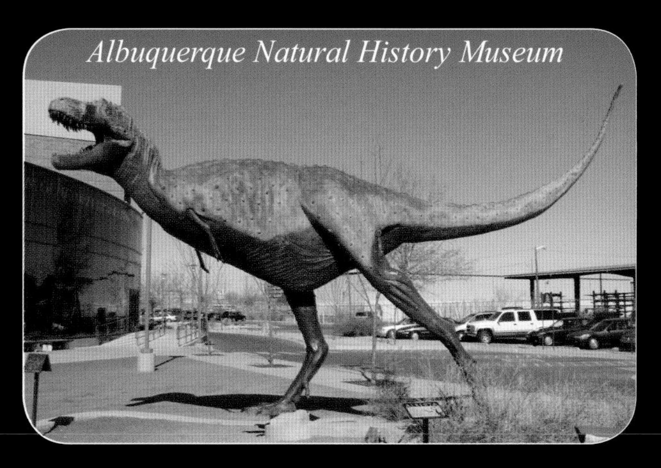

Albuquerque Natural History Museum

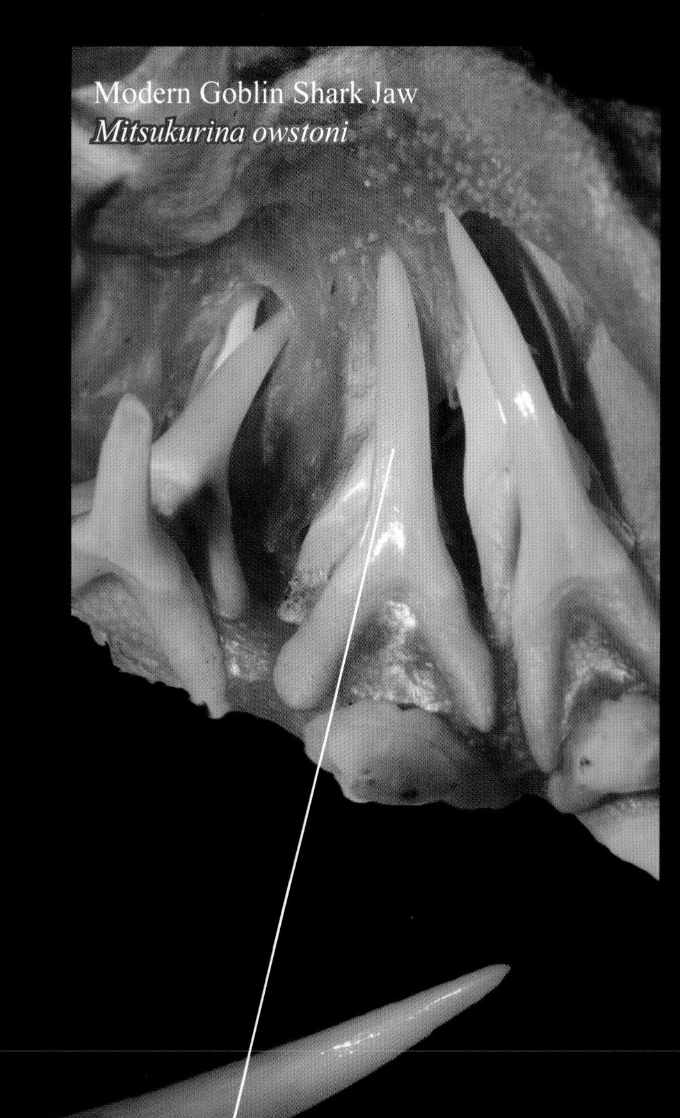

Modern Goblin Shark Jaw
Mitsukurina owstoni

Modern Goblin Shark Tooth
Mitsukurina owstoni

Dinosaur-Era Shark Tooth
Scapanorynchus (species undetermined)
New Mexico Museum of Natural History and Science, Albuquerque, USA

Sharks: Living with Both Dinosaurs and Humans

A shark has many differently shaped teeth, just like humans. Compare these fossil teeth to those of a living shark, and notice their similarity.

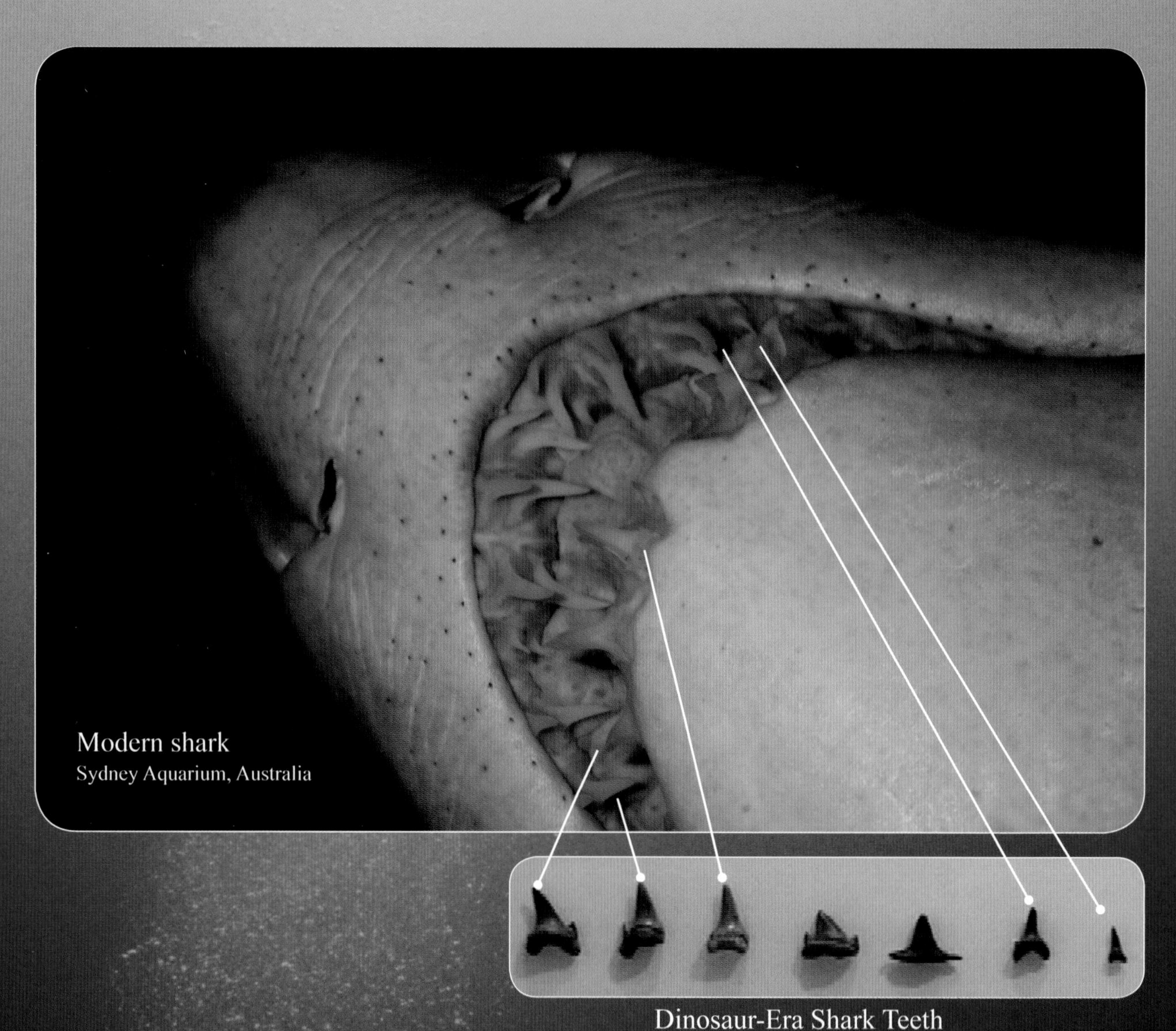

Modern shark
Sydney Aquarium, Australia

Dinosaur-Era Shark Teeth
Unidentified species
Cretaceous, Alberta, Canada
Dinosaur Provincial Park, Alberta, Canada

Summary

The theory of evolution states that as time passed, new types of fish evolved. According to the idea of the survival of the fittest, these bigger and better fish overpowered and outcompeted their predecessors. If evolution is taken at face value, then the sharks and rays should have dramatically changed over time, from one type into another. But in these particular examples of cartilaginous fish, I see very little difference between ancient and modern forms.

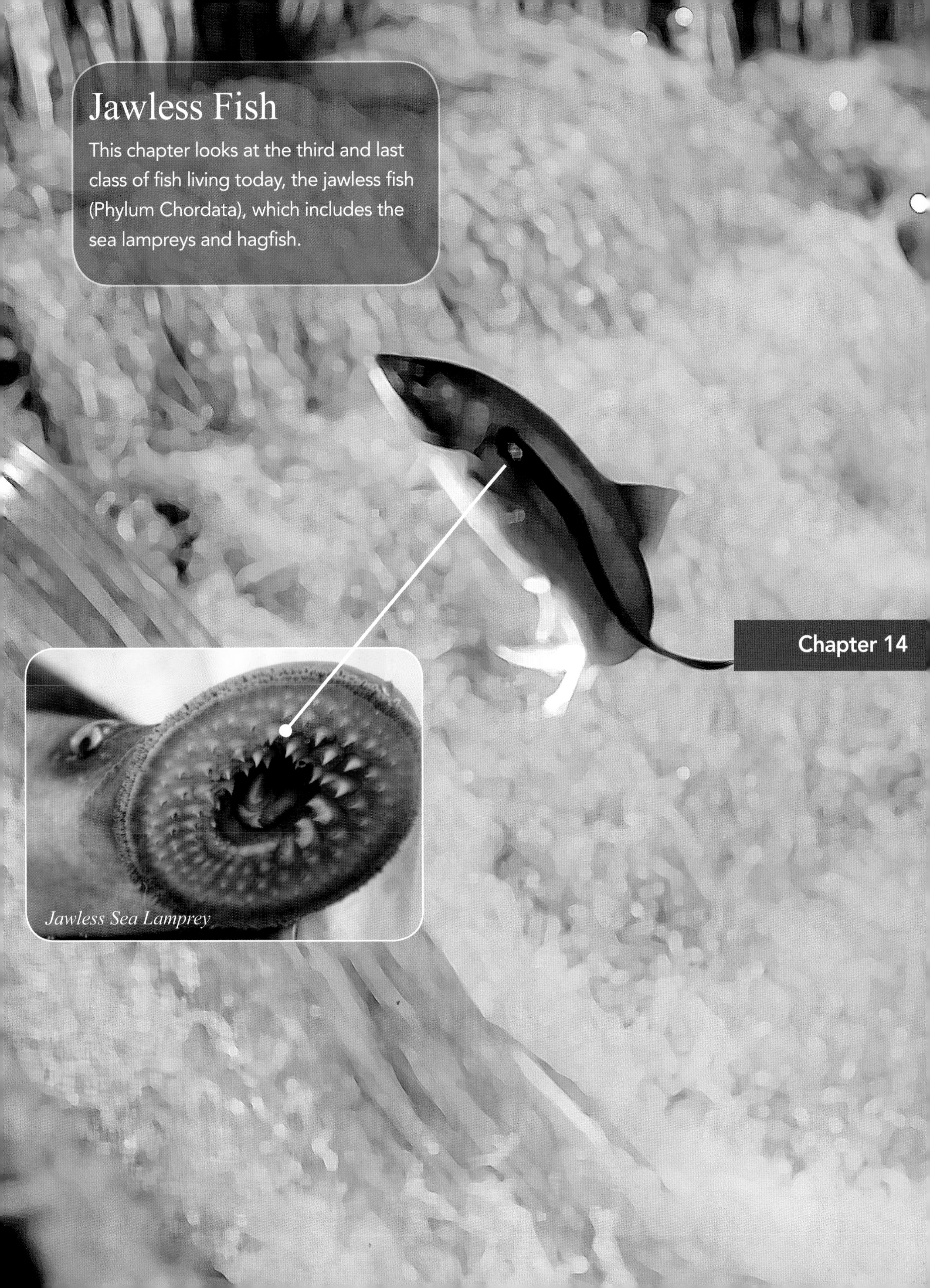

Jawless Fish

This chapter looks at the third and last class of fish living today, the jawless fish (Phylum Chordata), which includes the sea lampreys and hagfish.

Chapter 14

Jawless Sea Lamprey

"Jawless fish are still alive today and are nearly unchanged."

Jawless Fish — "Almost Unchanged"

According to the theory of evolution, jawless fish are the oldest type of fish and they theoretically evolved into cartilaginous sharks and rays which then evolved into bony fish. If jawless fish are the evolutionary predecessors of other fish, one would expect them to be extinct, according to the idea of the survival of the fittest.

While in Australia, we interviewed Dr. John Long, head of science at the Museum Victoria in Melbourne and curator of vertebrate paleontology at the Western Australian Museum. Dr. Long, who has written a textbook on the evolution of fish, is considered an expert on this topic. He contends that the jawless fish alive today are essentially unchanged from those living during the time of the dinosaurs. Again, this statement supports my hypothesis that if evolution is not true, I should find modern jawless fish in dinosaur rock layers.

"*The modern* [jawless] *lamprey and hagfish have fossil records spanning back to the Carboniferous Period* [which is even older than the dinosaur layers], *remaining almost unchanged.*"[1]

— Dr. Long

Dr. John Long, head of science at Museum Victoria in Melbourne, Australia, is the author of ***The Rise of Fishes: 500 Million Years of Evolution.***

Museum Victoria, Australia

What Do You Think?

Lamprey

Hagfish

Above: *Jawless hagfish are alive today and were living at the time of the dinosaurs.*

Summary of Fish Sections

My experiment predicted I would find examples of modern-appearing fish with dinosaurs if evolution was not true. I found not only bony fish, but also cartilaginous fish and jawless fish — examples from all three fish groups living today. Superficially, these look like they could be the same types of animals. Some may say, "*So, you may have proved your point in regard to the fish groups living today, but what about the land vertebrates? Surely modern types of land animals have not been found in dinosaur layers, and your evolution test will fall apart here.*" To this, I respond, "*Read on.*" I have found examples from *every type* of land vertebrate living today: amphibians, reptiles, birds, and mammals. The fossils we found were not strange, unrecognizable animals, but modern-appearing land vertebrates.

Do you think these findings are significant or insignificant?

Amphibians

This chapter deals with the two major types of amphibians (Phylum Chordata) living today, frogs and salamanders.

Chapter 15

"I never imagined a Tyrannosaurus rex stepping on a salamander or a frog, yet this is the fossil reality."

Salamander Found in Belgian Coal Mine with 29 Dinos

Debbie and I toured some of the great dig sites of Europe for this project. On one trip, we traveled to Bernissart, Belgium, where the fossilized remains of 29 *Iguanodon* dinosaurs were found in a mine in 1878. These dinosaurs were carefully removed and are now preserved under glass at The Royal Belgian Institute of Natural Sciences in Brussels. At this museum we viewed the plants and animals found with these dinos, and came across our first modern-appearing fossilized amphibian, a salamander.

Dinosaur-Era Salamander
Species Unidentified
Cretaceous, Bernissart, Belgium
The Royal Belgian Institute of Natural Sciences, Belgium

Modern Salamander
Species Unidentified

Belgian Dinosaurs

Clockwise from top left: **A:** Author meeting with the curator of The Iguanodon Museum, in Bernissart, Belgium, Mr. A. Nee. **B:** Curator of Museum standing on top of original opening of coal mine where dinosaurs were found in 1878. **C, D:** Museum painting and drawing of *Iguanodon* dinosaurs found in coal mine. **E:** Foot bones of *Iguanodon* dinosaur on display in Bernissart, Belgium. **F:** Museum sign. **G:** Original dinosaur bones on display under glass at Royal Belgian Institute of Natural Sciences in Brussels, Belgium. **H, I:** Belgian miners at Bernissart who worked in the mine where the fossil salamander was found.

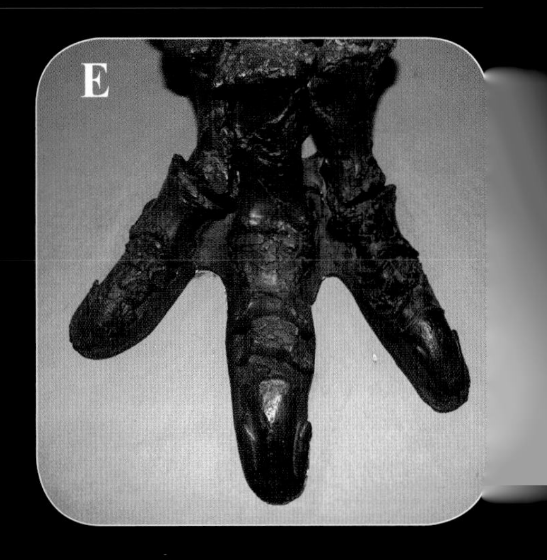

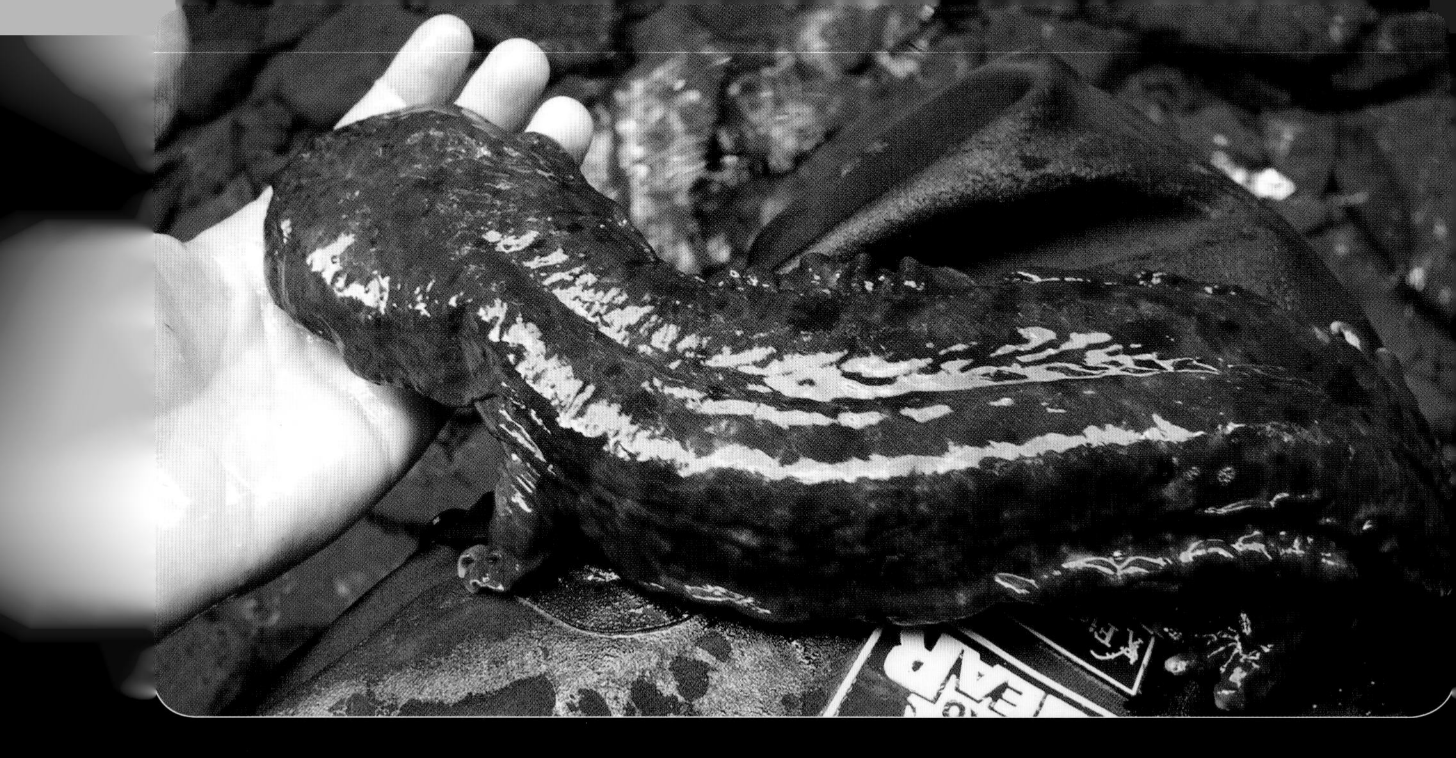

Hellbenders

Hellbenders are large carnivorous salamanders that live today in the streams of the Ozark and Appalachian mountain ranges. Most people have never seen one. I certainly have not, even though I have fished these areas fairly extensively.

One day, at a museum outing in Australia, I photographed a hellbender skeleton (middle of next page). It reminded me of the dinosaur-era amphibian *Karaurus, which* I had encountered in museums over the years (bottom of next page). I had always thought *Karaurus* to be an unusual, bizarre, extinct amphibian. But after comparing a modern hellbender skeleton to these fossils, I'm not so sure, especially in light of the large variations in the shape of the two fossil *Karaurus specimens on the next page.*

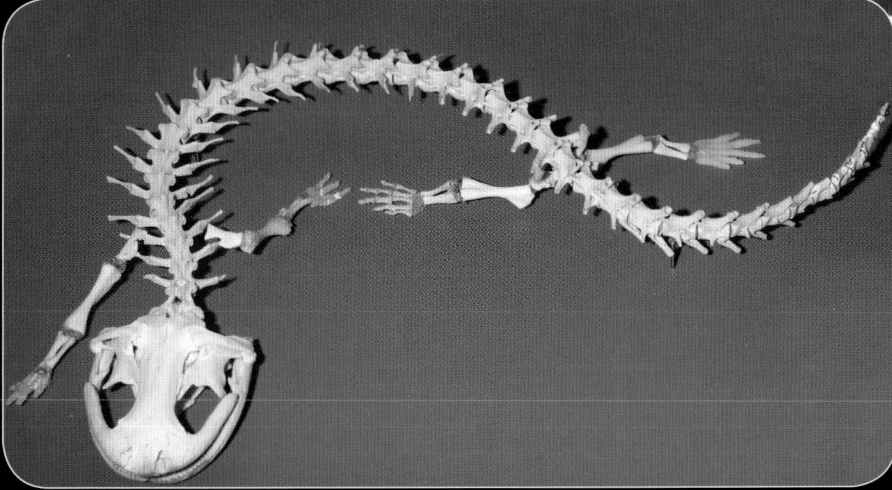

Modern Hellbender
Cryptobranchus alleghenіensis
South Australian Museum, Adelaide

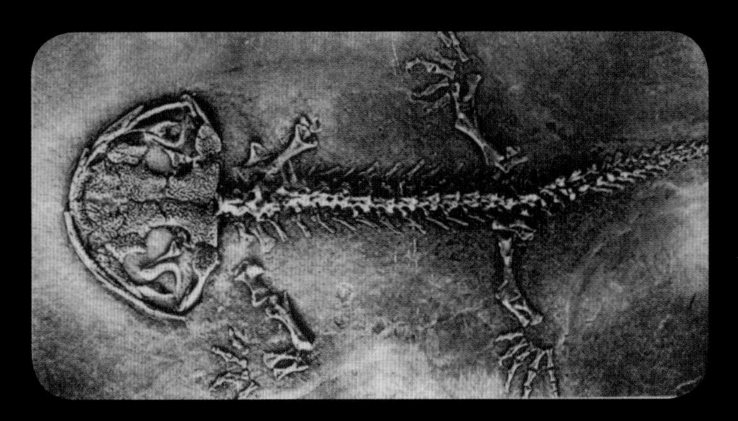

Dinosaur-Era Salamanders
Two fossils of *Karaurus*
Karaurus (species undetermined)
Jurassic
Redpath Museum, Montreal, Canada

Frogs

The other major group of amphibians living today are the frogs and toads. Were they also alive at the time of the dinosaurs? At the Redpath Museum in Montreal where Dr. Robert Carroll, an expert on fossil amphibians, is the curator, I saw a display that answered this question. Here was a drawing of a dinosaur-era fossil frog that was reminiscent of a common frog living today.

Red-eyed Tree Frog
Agalychnis ***calidryas***

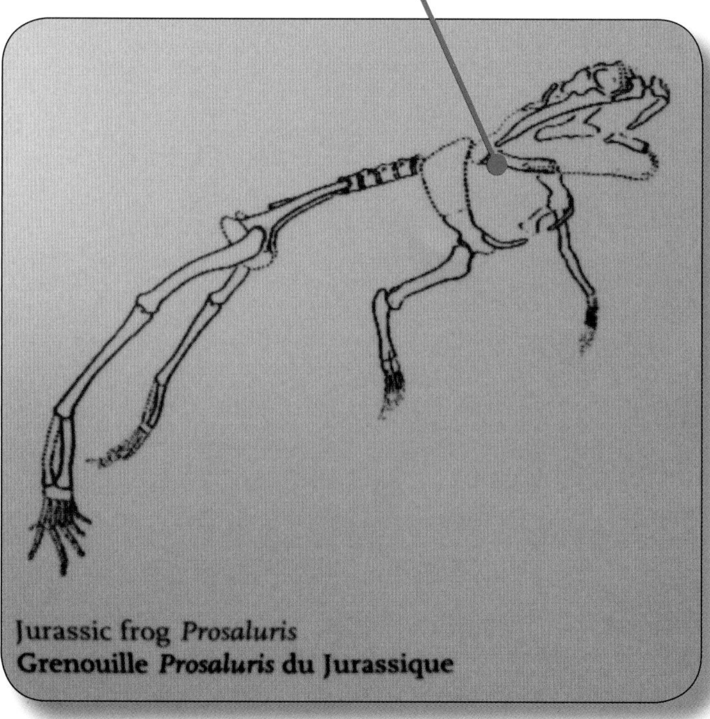

Museum Drawing Dinosaur-Era Frog
Prosaluris ***(species unidentified)***
Jurassic
Redpath Museum, Montreal, Canada

Redpath Museum, Canada

Above: *Dr. Robert Carroll (left), curator of vertebrate paleontology and professor of zoology at the Redpath Museum in Montreal, standing alongside author.*

Above: *Redpath Museum in Montreal.*

Below: *Dinosaur-era frogs and salamanders on display on the far wall, behind this dinosaur.*

Above: *Salamanders lived alongside dinosaurs.*

"During the Mesozoic Era, the age of dinosaurs, there were many animals that would be recognizable by people today...There would have been soft-shelled turtles, garfish swimming around at that time, lizards, frogs, salamanders, all very similar to the forms we see today."[2]

— Dr. Breithaupt

Dr. Brint Breithaupt, director, University of Wyoming Geological Museum.

Summary

Both major amphibian groups living today, frogs and salamanders, have been found in ancient rock layers and appear relatively unchanged, supporting my hypothesis that if evolution was not true, I would find modern types of animals in dinosaur layers. [3]

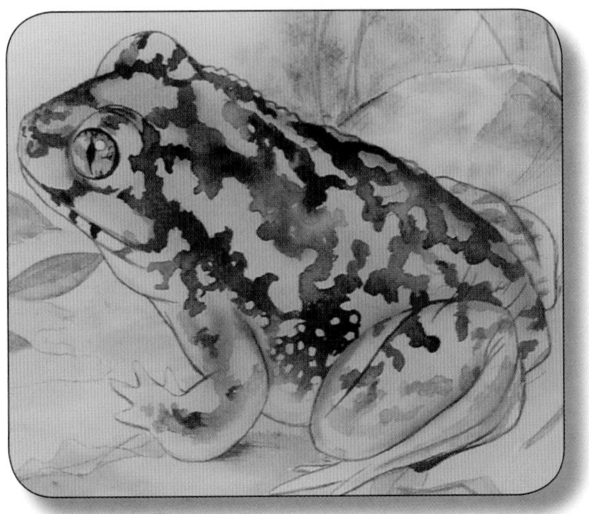

Dinosaur-era survivor

What Do You Think?

Crocodilians

This chapter deals with crocodilians, one of the four reptile groups living today (Phylum Chordata). This group consists of alligators, crocodiles, and gavials.

Chapter 16

"The alligators found at Dinosaur Provincial Park look like those living today."

Comparing Modern and Dinosaur-Era Alligators

Alligators today live along the Gulf Coast of the United States. They have broader snouts than crocodiles, and their teeth do not make a notch or indentation in the upper jaw. I have been able to film live alligators in Louisiana and at many zoos around the world.

When we filmed at Dinosaur Provincial Park in 1997, there was a fossilized alligator skull on display which had been found in the dinosaur layers there (bottom of next page). When I saw it, I wondered if it could be similar to the alligators living today in Louisiana, but I was not sure — it had a name quite dissimilar to the American alligator.

Getting an aerial view of a live alligator to compare to a fossil skull is a little tricky, as you can imagine, but thankfully, the juvenile alligator at the World Aquarium in St. Louis allowed me to stand directly over its head (below, left). Later, at the Redpath Museum in Montreal, I was also able to take an aerial view photograph of a modern alligator *skull*.

The two skulls on the next page, a fossil alligator skull from Dinosaur Provincial Park and a modern alligator skull, look very similar! Wouldn't you agree? (The fragile nasal septum in the fossil skull and some of the other ridges were presumably not preserved.)

Live Alligator
Alligator mississipiensis
World Aquarium, Missouri, USA

Modern Alligator
Alligator mississipiensis
Nebraska State Museum, Lincoln, USA

Modern Alligator Skull
Alligator mississipiensis
Redpath Museum, Montreal, Canada

Dinosaur-Era Alligator
Nasal septum not preserved
Albertochampsa langstoni
Cretaceous, Alberta, Canada
Dinosaur Provincial Park, Alberta, Canada [1]

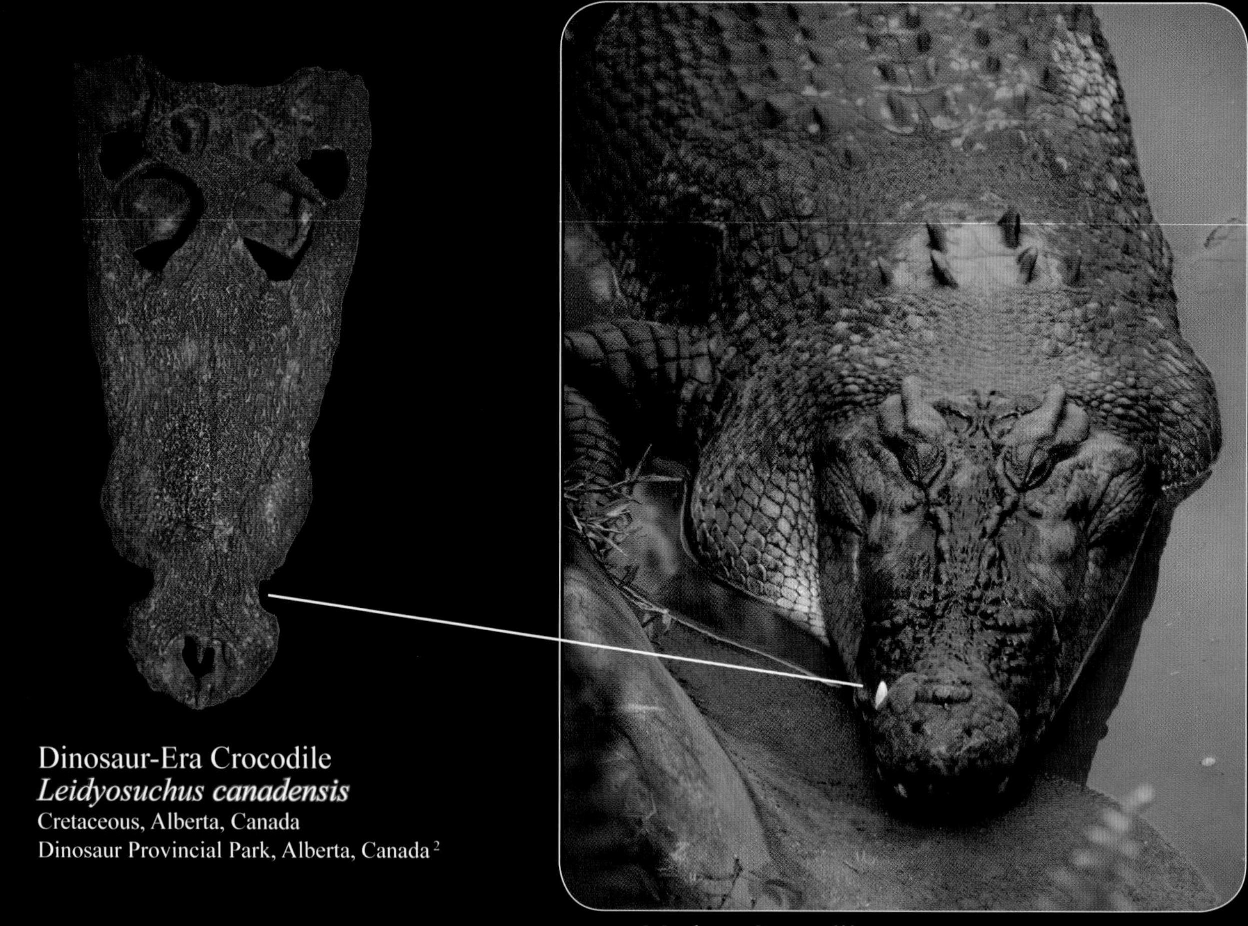

Dinosaur-Era Crocodile
Leidyosuchus canadensis
Cretaceous, Alberta, Canada
Dinosaur Provincial Park, Alberta, Canada [2]

Modern Crocodile
Crocodylus porosus
Australia

Crocodiles

Above is a fossil croc skull found at Dinosaur Provincial Park in Alberta, Canada, the same location where the fossil alligator was found. Compare this fossil skull to this modern Saltwater Crocodile we photographed in Australia (above, right). The fossil crocodile does not have the lower jaw attached, to show the lower teeth protruding upwards, but they can be inferred by the notch in the upper jaw. At the very least, I conclude that crocs are living today and lived with dinosaurs.

Dinosaur-Era Gavial
Steneosaurus bollensis
Jurassic, Germany
Carnegie Museum of Natural History, USA

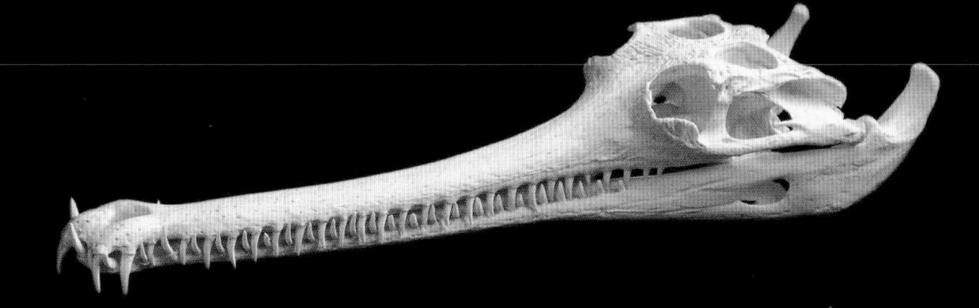

Modern Gavial Skull
Gavialis gengeticus

Gavials

At the Carnegie Museum, I saw this odd-looking fossil reptile (top of page) the likes of which I had never seen before. Its snout was very narrow, almost like a rod, yet the rest of its body looked like a crocodile. Its name was strange too, *Steneosaurus bollensis*. The sign at the museum suggested that this crocodilian was "similar to the modern needle-nosed gavial." Since I was unfamiliar with gavials which today live in Bangladesh, India, Pakistan, and Burma, I had to wait until I got home to find a photo of one.

When I placed the image of a modern gavial skull alongside this dinosaur-age "gavial-like" crocodile, they looked incredibly...similar. Now compare their genus names in blue.

Summary

All three families of crocodilians living today, crocodiles, alligators, and gavials, have been found in dinosaur rock layers and appear basically unchanged. Amazing.

Modern Gavial
Gavialis gengeticus

Snakes

This chapter addresses modern types of snakes (Phylum Chordata).

Chapter 17

"I had never heard of dinosaurs living alongside boa constrictors, but they did. It is a fact."

Boa Constrictor Found at Hell Creek Montana

When you think of dinosaurs, do you also think of snakes? I certainly did not. Dinosaurs, I thought, were ancient reptiles, and snakes were modern reptiles.

At the Hell Creek dinosaur dig site in Montana, scientists found not only a *Triceratops* and a T*yrannosaurus rex*, but also fossil bones of a boa constrictor. Now this may seem insignificant to you, but for me it was *very* significant — another reptile group to test evolution. A reconstructed model of this dinosaur-age boa constrictor from Montana is on display at the Milwaukee County Museum. (See Page 146.)

Top: *Museum display showing dig site at Hell Creek where dinosaurs were found along with a boa constrictor.*

Bottom: *Modern boa constrictor wrapping itself around and preparing to eat a rodent.*

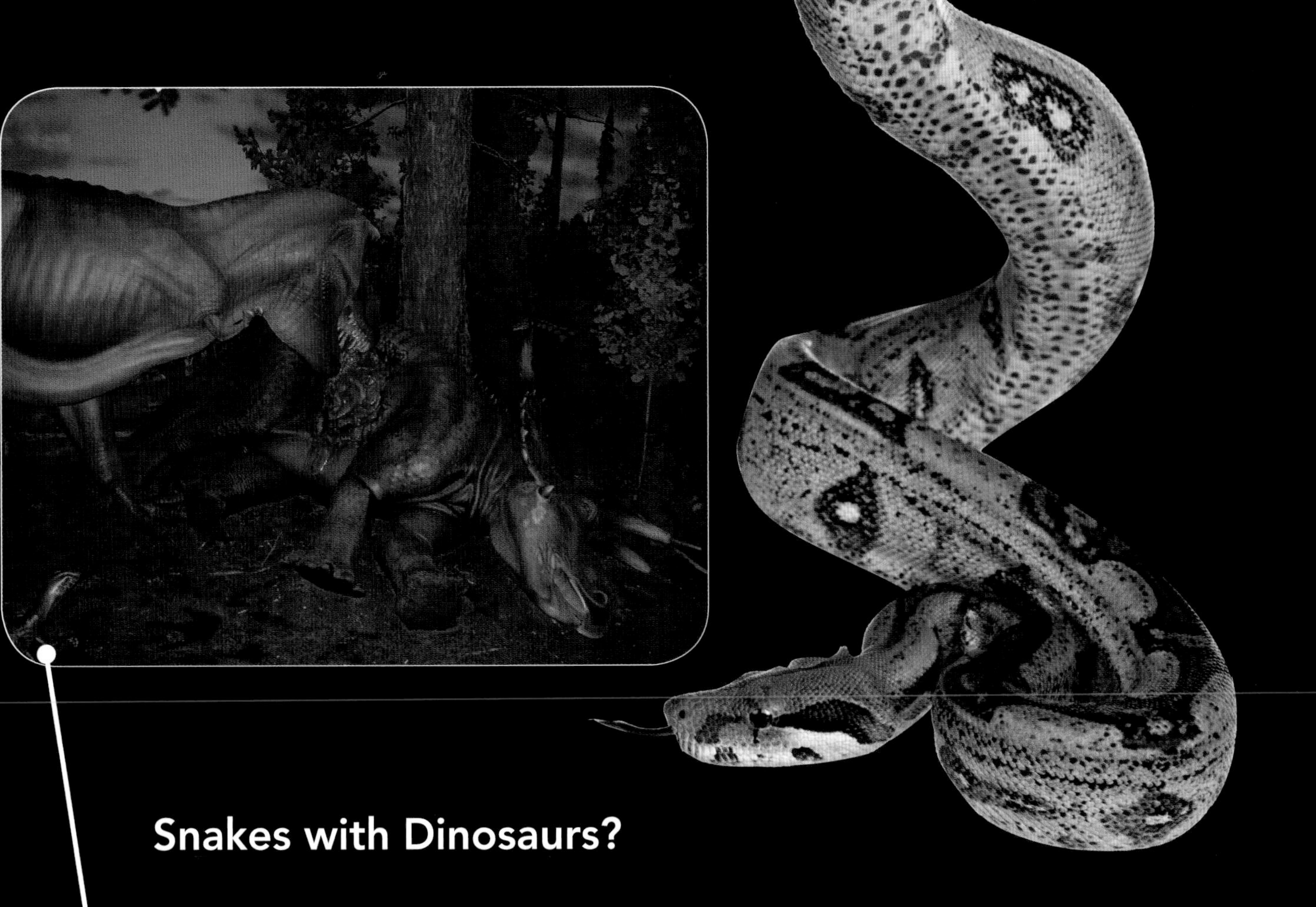

Snakes with Dinosaurs?

Above: *Museum display showing what life may have been like at Hell Creek, Montana, where a fossil boa constrictor was found. In this museum diorama, T. rex takes a bite out of the side of a Triceratops while two small Dromaeosaurus dinosaurs look on.*

Below: *Close-up photo of Dromaeosaurus dinosaurs.*

Museum chose not to display the fossil bones of the boa constrictor; rather, they reconstructed what the snake looked like, based on the fossil evidence. When I placed this reconstructed dinosaur-era boa constrictor (right) next to a photograph of a live boa constrictor (below, right), they looked like the same snake. Remember, scientists cannot tell the color of an animal when they find fossil bones. In looking at any reconstruction of an animal, you have to ignore the parts that have been interpreted by the artist.*

Museum Reconstruction
Dinosaur-Era Boa Constrictor
Cretaceous, Hell Creek Formation, Montana
Milwaukee Public Museum, USA

Other Snakes

Besides this fossil boa, fossil snakes have also been found in Cretaceous dinosaur fossil layers in India,[1] the United States,[2] Madagascar,[3] Niger,[3] Sudan,[3] Spain,[3] Argentina,[3] Romania,[3] and Algeria.[4]

Living Boa Constrictor
Reptile Gardens, Branson, Missouri, USA

Summary

Modern-appearing snakes have been found in dinosaur layers all over the world, a further evidence for my experiment.

***See Appendix D:** *Assumptions in Fossil Reconstructions — Color*

Lizards

This chapter deals with the third type of modern reptiles, lizards (Phylum Chordata).

Chapter 18

"I thought lizards were modern reptiles and dinosaurs were ancient reptiles, but the fossils proved me wrong."

Dinosaur-Era Iguana-Like Lizard
Polyglyphanodon sternbergeri
Cretaceous, Utah, USA
Carnegie Museum of Natural History, USA

Iguana-Like Lizard at the Carnegie Museum

The fossil lizard above is the same animal depicted in the drawing at the top of the next page. The Carnegie Museum display says that this fossil lizard "resembles the living iguanas but is distinguished by its teeth — these are widened side to side and have sharp cutting edges."

Now look at the teeth of the modern iguana on the next page. Did the scientists assign this new genus name to the fossil simply because its teeth were wider and sharper than the living iguana? I will not venture to guess but from the anatomy that I can see, I will say these lizards look as though they could be siblings.

Dinosaur-Era Iguana-Like Lizard
Museum Drawing
Polyglyphanodon sternbergeri
Cretaceous, Utah, USA
Carnegie Museum of Natural History, USA

Modern Iguana
Modern Iguana Teeth
Iguana iguana
Cozumel, Mexico

Ground Lizards

Next, compare this modern ground lizard to the Jurassic lizard from Germany on this page. Look carefully at the number of toes, the length of the toes, the length of the ribs, the width of the tail vertebrae, etc. They look very similar to me.

Modern Ground Lizard
South Australian Museum, Adelaide

Dinosaur-Era Lizard
Jurassic, Solnhofen, Germany
Jura Museum, Germany

At Solnhofen, Germany, where the famous fossil bird *Archaeopteryx* was found along with a dinosaur, quarry workers also found this fossil lizard (below) that — according to evolution scientists — looked very similar to a modern tuatara. [1]

Dinosaur-Era Tuatara-Like Lizard
Homeosaurus maxmilliani
Jurassic, Eichstatt, Germany
Harvard Museum of Paleontology, USA

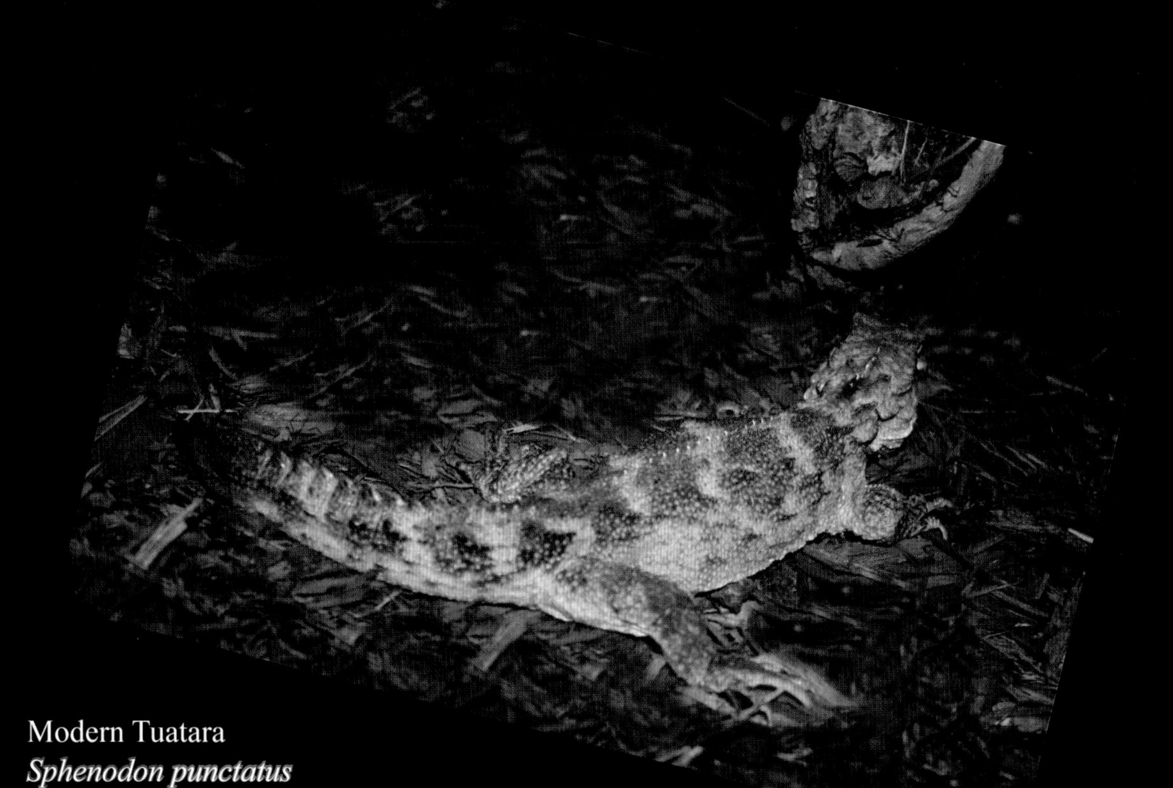

Modern Tuatara
Sphenodon punctatus

Gliding Lizards

The most interesting fossil I found from the dinosaur fossil layers was a *gliding* lizard on display at the Milwaukee Public Museum in Wisconsin (below). Compare this fossil to a living gliding lizard on the next page. Little has changed with this type of lizard, including the elongated rib bones which act as struts to support the membranous "wings." Now compare the genus names in blue.

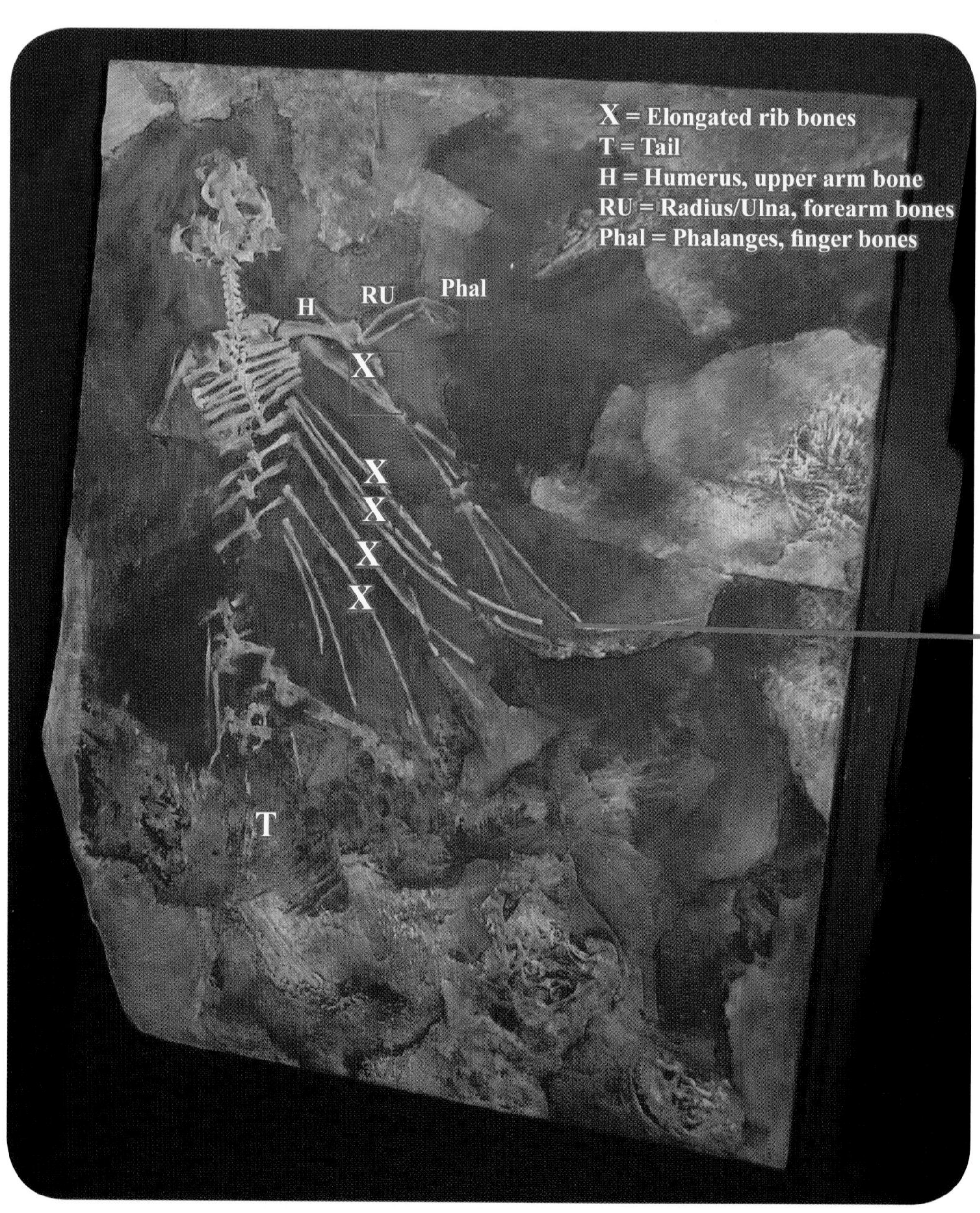

Dinosaur-Era Gliding Lizard
***Icarosaurus** siefkeri*
Triassic, New Jersey, USA
Milwaukee Public Museum, USA

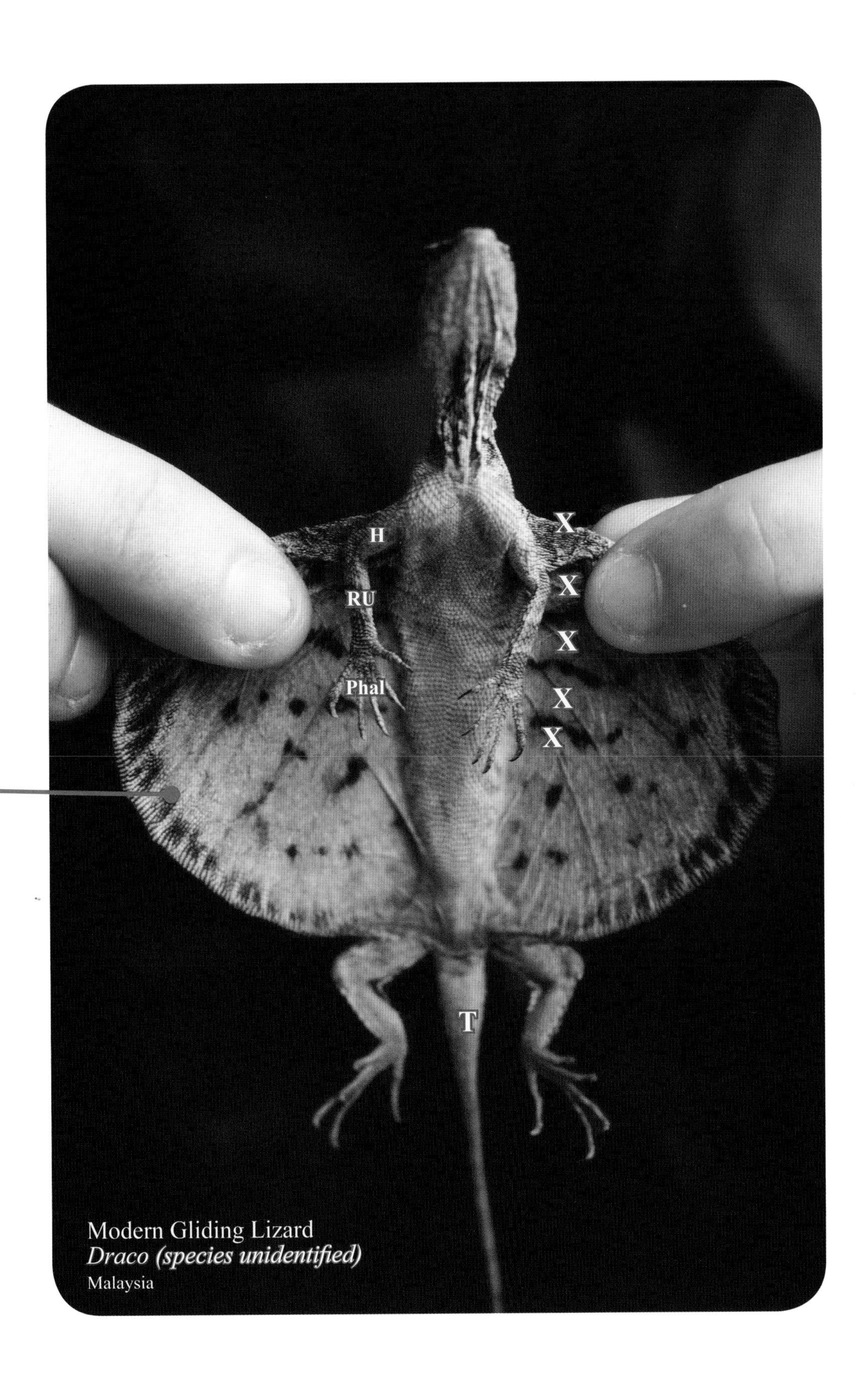

Modern Gliding Lizard
Draco (species unidentified)
Malaysia

Modern Gliding Lizard
***Draco** (species unidentified)*
Malaysia

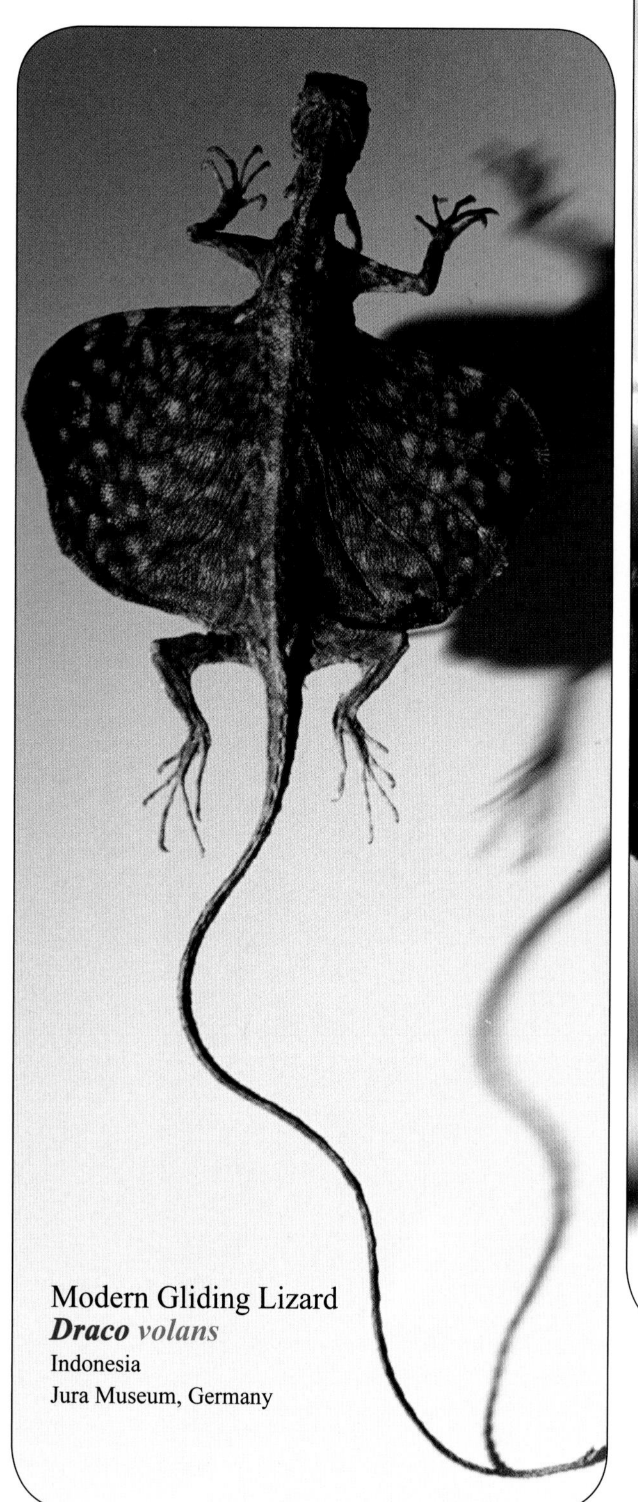

Modern Gliding Lizard
***Draco** volans*
Indonesia
Jura Museum, Germany

Summary

Many modern-appearing lizards have been found alongside dinosaurs, including iguana-like lizards, ground lizards, tuatara-like lizards, and gliding lizards.

Turtles

This chapter deals with the fourth and last type of modern reptiles, turtles (Phylum Chordata).

Chapter 19

"I was unaware that so many modern-appearing turtles were alive during the time of the dinosaurs."

Box Turtles

I had always played with box turtles as a kid. We used to catch them in the woods, put them in a cardboard box, feed them some lettuce, and try to make them drink out of a jar lid. Later, my parents would make us release them back into the wild. Because of my childhood play, I was very familiar with their shells.

As an adult, nearly 30 years later, I saw a fossil box turtle at the Milwaukee Public Museum that looked nearly identical to the modern form. I must admit I was a little shocked. Compare this fossil turtle on the left with the modern box turtle below. Look closely at the number of toes and the shell pattern.

Dinosaur-Era Turtle
Museum reconstruction
Plesiobaena antiqua
Cretaceous, Hell Creek, Montana, USA
Milwaukee Public Museum, USA[1]

Modern Eastern Box Turtle
Terrapene carolina
Missouri, USA

Modern Slider Turtle
Trachemys scripta
Louisiana, USA[2]

Dinosaur-Era Turtle
Plesiochelys (unidentified species)
Jurassic, Solnhofen, Germany
Carnegie Museum Of Natural History, USA

Jurassic Turtles

This fossil turtle (above, right) was found in Jurassic fossil layers. Compare the fossil turtle with a modern pond turtle (above, left). Modern-appearing turtles lived at the same time as the dinosaur Camptosaurus (below, left) and the long-necked dinosaurs (below, right).

The Two Types of Turtles

There are only two sub-orders of turtles living today, the side-neck turtles (Pleurodira) and the hidden-neck turtles (Cryptodira).[3] Both of these types of turtles were also alive during the time of the dinosaurs.[4,5] Side-necked turtles, such as the living Australian Snake-Necked Turtle (below) and the dinosaur era turtle *Proterochersis robusta,* tuck their head partially *around the side of their body* when they close their shell.[4,6] Hidden-neck turtles, such as the modern box turtle, the modern slider turtle, and the dinosaur-era turtle *Solnhofia parsoni*, pull their neck inside the shell, *between* the shoulders.[5,7,8]

Right: *At Dinosaur National Monument, they found a turtle in the same layers as the dinosaur Camarasaurus.*

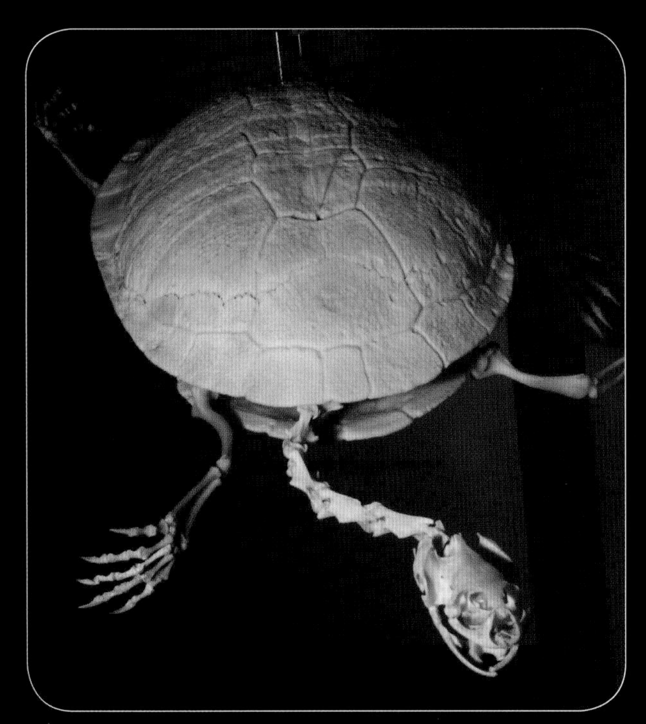

Australian Snake-Necked Turtle
Modern Side-neck Turtle
Chelodina longicollis
South Australian Museum, Adelaide

Dinosaur-Era Hidden-Neck Turtle
Glyptops plicatulus
Dinosaur National Monument, Utah, USA

Turtles Found at Dinosaur National Monument

Above: *Worker removing rock from around a turtle shell found at Dinosaur National Monument, Utah, USA*

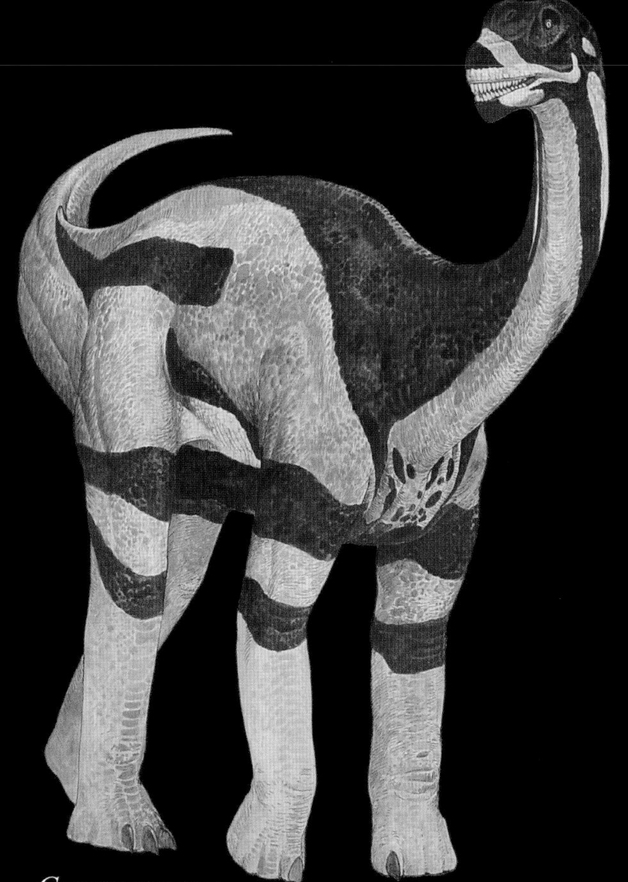

Camarasaurus
Jurassic, Dinosaur National Monument
Vernal, Utah, USA

Above: *Park ranger at Dinosaur National Monument pointing to a dinosaur leg bone, which remains in the wall.*

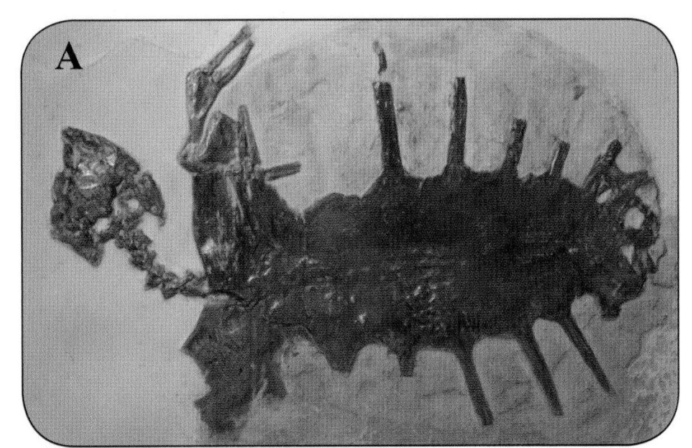

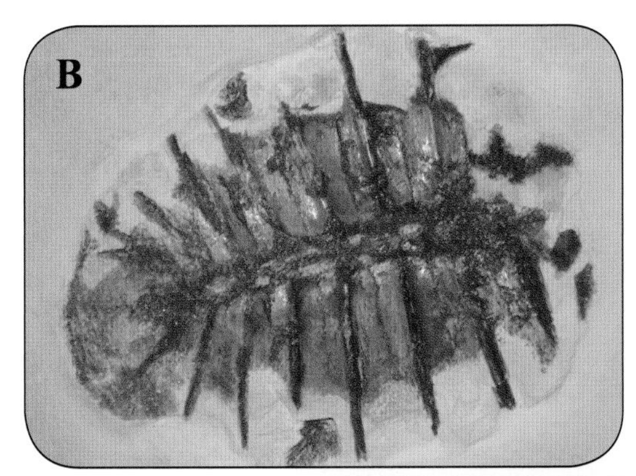

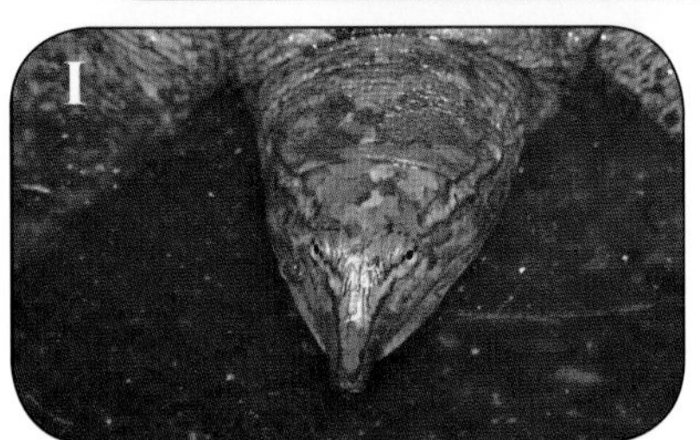

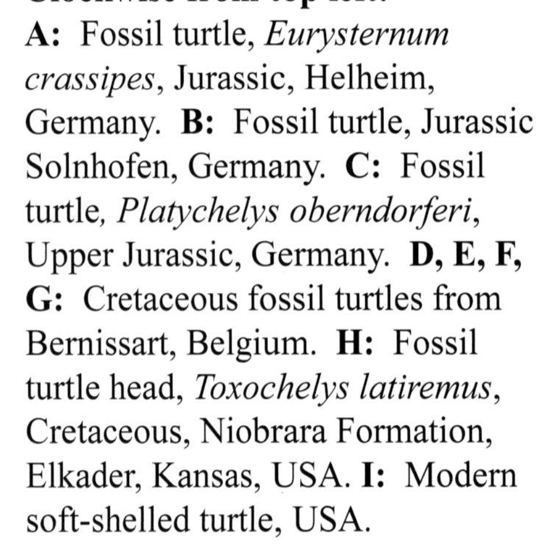

Other Dinosaur-Era Turtles

Clockwise from top left:
A: Fossil turtle, *Eurysternum crassipes*, Jurassic, Helheim, Germany. **B:** Fossil turtle, Jurassic Solnhofen, Germany. **C:** Fossil turtle, *Platychelys oberndorferi*, Upper Jurassic, Germany. **D, E, F, G:** Cretaceous fossil turtles from Bernissart, Belgium. **H:** Fossil turtle head, *Toxochelys latiremus*, Cretaceous, Niobrara Formation, Elkader, Kansas, USA. **I:** Modern soft-shelled turtle, USA.

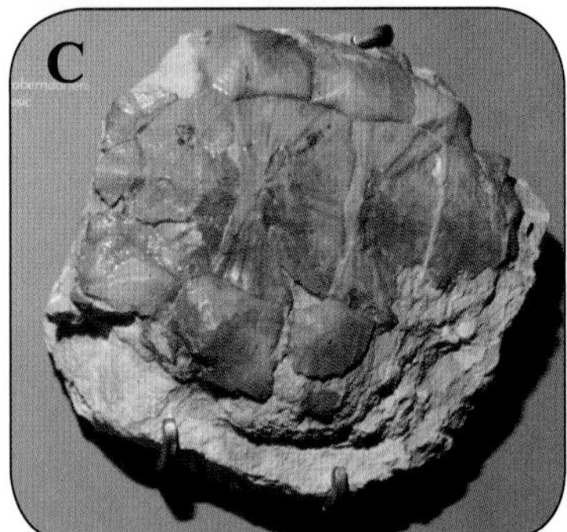

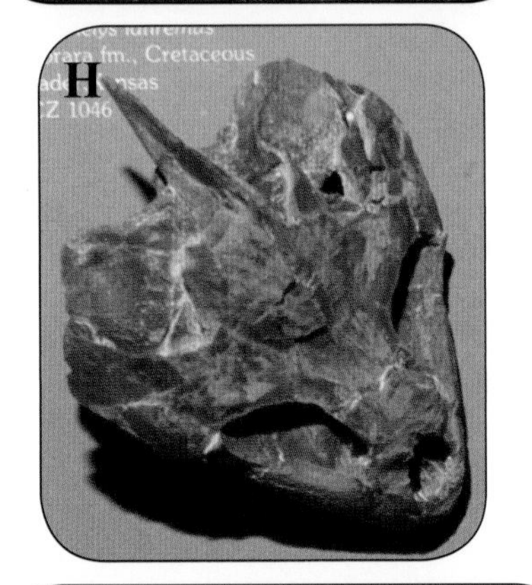

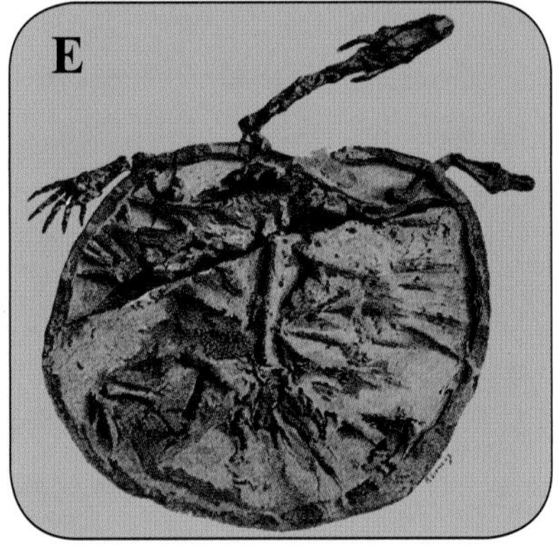

Summary: Turtles and Reptiles

Both sub-orders of turtles living today, the side-neck turtles (Pleurodira) and the hidden-neck turtles (Cryptodira) were also alive during the time of the dinosaurs. [4, 5]

All four orders of reptiles living today: Crocodilia, Sphenodontia (tuataras), Squamata (snakes and lizards), Testudines (turtles) have been in dinosaur rock layers and they look nearly the same. Amazing! Now, on to birds.

Birds

This chapter address modern types of birds (Phylum Chordata).

Chapter 20

"I thought that only extinct, toothed birds, like Archaeopteryx, were alive during the time of the dinosaurs."

Were Modern Types of Birds Living with Dinosaurs?

If you asked visitors leaving a natural history museum this question, "Did *modern* types of birds live at the same time as the dinosaurs?" what response do you think you would get? Most likely, you would get a resounding "No!" Before I began the active phase of my experiment in 1997, I had this same impression, that only extinct, unusual birds were present during the time of the dinosaurs, such as the toothed bird *Archaeopteryx*. In the 60 museums I visited, not once did I see a single fossil of a modern bird from a dinosaur layer. (I did see a reconstruction of a fossil bird in Milwaukee, but not the actual fossil.)

My experiment predicted that if evolution was not true, I would find modern birds with dinosaurs. I already had fairly robust evidence that evolution was not true (animals did not change significantly over time) for the other animals such as reptiles, amphibians, fish, worms, sponges, shellfish, crustaceans, insects and echinoderms, but birds seemed to be different. Why?

It turns out that they are not different rather, they are simply not *displayed* in the museums we visited. This came to light when I interviewed the paleontologists from 1997 to 2008.

As I mentioned earlier, Dr. William Clemens, a paleontologist at the University of California, Berkeley, got into a bitter argument, played out in the national press, in the late 1980s with the authors of the asteroid theory. (I interviewed scientists from both sides of this great scientific debate.) The asteroid theory suggested that dinosaurs went extinct because an asteroid hit the earth, resulting in an ecological disaster. The fire, smoke, and dust from the asteroid impact theoretically blocked off the sun, causing a cooling of the atmosphere. This climate change then killed off all of the dinosaurs, or so the theory goes.

Dr. Clemens simply did not believe the asteroid-extinction theory because, in his opinion, there was a lack of evidence to support it. He had never found the smoking gun — a field of dead dinosaurs that coincided with an asteroid impact. Because of this, he and his colleagues set out to prove the theory wrong.

Dr. Clemens reasoned that if an asteroid struck the earth and killed off the dinosaurs by an ecological disaster, then the other animals living at that same time, such as amphibians, insects, and birds, should have gone extinct too. These animals are much more sensitive to environmental changes and pollutants and should have perished along with the dinosaurs.

He and his group began to look for amphibians, insects, and birds in the dinosaur layers. He reasoned that the more examples he found with the dinosaurs, the less likely it was that an asteroid impact had caused the extinction of the dinosaurs since these sensitive animals are still living today. This was the backdrop for our interview when I asked Dr. Clemens the million-dollar question: Had he found any *modern-appearing* birds with dinosaurs? Dr. Clemens recounted that a graduate student of his had found a parrot bone from the dinosaur layers. (When he said this, I almost fell out of my chair since his report fell in line with my original prediction about finding modern-appearing birds with dinosaurs.)

Dinosaur-Era Parrot

Dr. William Clemens, paleontologist, University of California, Berkeley.

"Paleontologists have been looking at these late Cretaceous collections and particularly focusing on the little scraps of bone. And what they're finding is, there is at that time, in the late Cretaceous, quite a diversity of modern groups of birds....Tom [Stidham, a graduate student at UC Berkeley] *has just had a little article published in* ***Nature***,[1] *in which he describes a late* ***Cretaceous parrot****. There are more pieces of information out there that are now being described in papers, that are being written or in review, showing that* ***other kinds of modern birds were present in the late Cretaceous****."*[2]

— Dr. Clemens

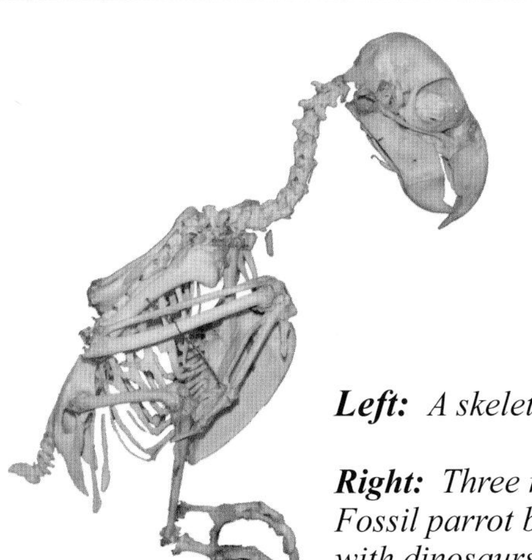

Left: *A skeleton of a modern parrot.*

Right: *Three modern baby Caique parrots. Fossil parrot bones have recently been found with dinosaurs.*

Flamingos, Cormorants, Sandpipers, and Dinosaurs

That same year I also interviewed Dr. Monroe Strickberger from the University of California, Berkeley, on the topic of proteins and the origin of life (for Volume I of this series). After the interview, I went back and read his 1996 college textbook, *Evolution*. What he said in his book about bird evolution raised my suspicions one more notch. I had to ask myself, "What was a flamingo doing with dinosaurs anyway?" In retrospect, I wonder why I had never seen any of these fossils displayed at the 60 museums we visited.

"Unfortunately, no feathered intermediates appear between Archaeopteryx and its dinosaur ancestors, nor do further birdlike fossils show up until about ten million years later in the Cretaceous period. ***These Cretaceous*** [dinosaur-age] ***fossils are exclusively those of aquatic birds or shore birds, a few already representative of modern groups such as flamingos, loons, cormorants, and sandpipers****, although some, such as Hesperornis, still retained reptile-like teeth."*[3]

— Dr. Strickberger

Dr. Monroe Strickberger, author of the college textbook ***Evolution****.*

Owls and Penguins

In February 1999, we interviewed Dr. Paul Sereno, from the University of Chicago on the topic of dinosaur evolution. During this interview, I asked Dr. Sereno my standard question about "asteroid survivors." His answer made me begin to feel that my experiment had been vindicated. He suggested that not only parrots but penguins and owls had been found in dinosaur rock layers too.

Dr. Paul Sereno, paleontologist, University of Chicago

*"What is becoming apparent is that many of the modern bird groups — **parrots**, maybe even **penguins**, and other kinds of groups like **owls** — evolved earlier in the dinosaur era, and we are beginning to pick up their traces."*[4]

— Dr. Sereno

Humboldt Penguin
Spheniscus humboldti
Milwaukee County Zoo, USA

Eastern Screech Owls
Megascops asio
Minnesota, USA

Dinosaur-Era Avocet
Museum Model
Hell Creek, Montana, USA
Milwaukee Public Museum

Avocet

Three years later, in February 2002, Debbie and I traveled to the Milwaukee Public Museum. Here I found yet another example of a modern type of bird found in dinosaur rock layers. The museum staff had found a modern-appearing avocet (left) along with a *T. rex* and a *Triceratops* dinosaur at Hell Creek, Montana. Avocets are living today in Louisiana and Texas.

The scientists at the Milwaukee Public Museum reconstructed a model of what this ancient avocet looked like, based on the fossilized bones. The reconstructed model is at the top of this page. Compare it to the living avocet to the left. Ignoring *all* of the colors of the feathers, which are not preserved in the fossils, including the yellow feathers around the eyes, do you think these two animals could be the same species?

Modern American Avocet
Recurvirostra americana

Albatross, Ducks, and Loons

Later, I came upon an interview with Tom Stidham, the discoverer of the fossil parrot that Dr. Clemens talked about. In this article, Dr. Stidham revealed several more birds from the dinosaur era. "*Until now, the only modern bird fossils uncovered from the Cretaceous* [dinosaur rock layers] *have been water birds: loons, duck-like waterfowl, shorebirds, and tube-nosed seabirds like the albatross.*"[5]

I asked myself: Why are these not on display at museums? How could these have been overlooked in the first 150 years of paleontology? What else is out there that has not been publicized?

Albatross

Duck

Loon with baby on back

Most or All Major Modern Bird Groups Lived with T. rex

Left: *Tyrannosaurus rex.*

Nature magazine is one of the most prestigious science magazines on the planet. When I pulled Tom Stidham's original parrot article from *Nature*, I learned something even more significant. The article suggested that *most* or *all* of the modern bird groups were alive during the time of the dinosaurs, not just owls, flamingos, sandpipers, penguins, cormorants, avocets, loons, albatrosses, and ducks. Now, this is truly amazing. Stidham wrote: "*The existence of this fossil* [parrot] *supports the hypothesis, based on molecular divergence data, that* ***most or all of the major modern bird groups were present in the Cretaceous.***"[1]

T. rex Eats *Triceratops*
Carnegie Museum of Natural History, USA

Which of These Two Museum Paintings Is More Accurate?

Shore Birds with ***Pterodactylus***
Geological Museum
University of Wisconsin, Madison, USA

Above: Museum artists frequently leave out modern types of birds when they paint images of dinosaur life. **Right:** This museum display got it right showing modern-appearing shore birds with a flying reptile. (Flying reptiles lived at the same time as dinosaurs.) **Below:** This modern shore bird looks similar to the ones pictured with the pterosaur (right).

Glaucous-winged Gull
Larus glaucescens
Alaska, USA

Summary

Based on fossil evidence, parrots, flamingos, cormorants, sandpipers, owls, penguins, avocets, and tube-nose albatross-like birds lived at the same time as dinosaurs. Based on molecular divergence data, most, or all, of the major modern bird groups were present during the time of the dinosaurs. For me, birds were a completely unexpected fulfillment of my hypothesis prediction and raised serious doubts about the theory of evolution.

This chapter discusses the three subclasses of mammals living today (Phylum Chordata) — placentals, pouched marsupials, and egg-laying monotremes.

Chapter 21

"In over 60 museums, I have only seen three dinosaur-era mammal skeletons, and not one of these was displayed with a modern mammal skeleton for comparison."

Dealing with Negative Evidence

Let's imagine that in 1970, someone else had carried out a similar investigation as mine. He or she may have *speculated* that if evolution was not true, one should find modern plants and animals in dinosaur rock layers. This hypothetical scientist may have found modern-appearing shellfish, crustaceans, and fish in dinosaur rock layers but not modern birds, mammals, or flowering plants, and suggested these fossils were still in the ground, as yet undiscovered. But the critics of that time would have capitalized on the *absence* of modern-appearing birds, mammals, and flowering plants and concluded that this scientist was dead wrong. Would these critics have been justified in their position? Would the absence of a group of modern plants or animals in certain rock layers prove that evolution is true?

Stegosaurus found at Dinosaur National Monument

The number of modern animals that were alive during the time of the dinosaurs has never changed; it is only our understanding of that reality that has changed as we have discovered more fossils.

It seems that every 15 years or so, whole new groups of organisms are added to the ever-growing list of modern-appearing plants and animals that have been found with the dinosaurs. Knowing this, can you make any firm conclusion about what type of animals or plants lived during the time of the dinosaurs?

The absence of fossil evidence is a tricky issue to deal with. Scientists agree that you can't read too much into the fact that you have not found a particular plant or a particular animal in any one rock layer. Dr. William Clemens, from the University of California, put this succinctly as he discussed the dangers of dealing with non-evidence.

"Now it's dangerous in paleontology to deal with negative evidence — the absence of things — because it could be so easily the result of non-preservation or you just didn't have the luck to find them." [1]

— Dr. Clemens

Dr. William Clemens, paleontologist, UC Berkeley.

Left: *Photograph of some of the scientists involved in the initial rush to discover dinosaur bones in the late 1800s. These scientists sometimes missed the great wealth of mammal bones alongside dinosaurs.*

The Last Frontier of Living Fossils: Modern Mammals with Dinosaurs

You might be thinking I am setting up this chapter to conclude that modern-appearing mammals have not been found in dinosaur rock layers. Actually, there has been great progress, but we are just getting started.

Before 1812, the year that mammal bones were first discovered in dinosaur rock layers, it was presumed by some scientists that mammals did not live during the time of the dinosaurs.[2] Even after this landmark discovery, scientists continued to refer to the time of the dinosaurs as "The Age of the Reptiles."

There were many reasons scientists continued to refer to the times when dinosaurs were alive as the "Age of the Reptiles." One reason was it was rare to find mammal bones in dinosaur rock layers. Another reason was the sheer size of the bones. Some dinosaur bones were four feet long and weighed hundreds of pounds, whereas many mammal bones were only inches long.

Left: Workers removing a large dinosaur leg bone using ropes and pulleys at Dinosaur National Monument in the early 20th century. Scientists easily spotted these large bones but only rarely found tiny mammal bone. Because of this disparity in fauna, they referred to the dinosaur era as the "Age of the Reptiles."

Left: *This* ***current museum display*** *is inaccurate. Mammals, have been found in all of the dinosaur rock layers.* [5]

The Age of Reptiles Is a Misnomer

It turns out that mammals are not all that rare in dinosaur rock layers. For example, in just one section of a Jurassic rock formation, 43 different species of mammals were found. [3] Nearly 300 ***genus groups*** of mammals have now been found in dinosaur rock layers and many of these genera have more than one species. Dinosaur-age mammals have been found in Europe, Canada, China, Australia, and the United States. Most of these are bits and pieces or single bones, but many complete skeletons have also been found. [4]

And as far as size, our perceptions are changing too. For a long time, scientists taught that the largest dinosaur-era mammal was about the size of a shrew or a mouse. With the discoveries of the 12-pound opossum-like *mammal* and a three-foot long, 30-pound collie-sized Tasmanian Devil-like mammal, dinosaur-era mammals are approaching modern mammal sizes. With all of this new information, the term "The Age of the Reptiles" is considered a misnomer. Unfortunately, some museums perpetuate this myth even today.

"In a sense, 'The Age of Dinosaurs' or 'The Age of Reptiles' for the Mesozoic is a ***misnomer****....Mammals are just one such important group that lived with the dinosaurs, coexisted with the dinosaurs, and survived the dinosaurs."* [6]

— Dr. Luo

Dr. Zhe-Xi Luo is curator of vertebrate paleontology and associate director of research and collections at the Carnegie Museum of Natural History in Pittsburgh.

Nearly 100 *Complete* Dinosaur-Era Mammal Skeletons Found

According to Dr. Zhe-Xi Luo at the Carnegie Museum of Natural History, nearly 100 *complete* skeletons of mammals have been found in dinosaur rock layers. This was surprising because I had only seen three of these skeletons in all of the museums we visited. It would be most helpful for my experiment if museums displayed these nearly 100 complete skeletons along with their closest modern relatives for comparison. Ah, life is never so easy when you are involved in a Grand Experiment.

Question: *"How many* ***complete*** *skeletons (of mammals) from the dinosaur era have been found?"*

Dr. Luo: *"So all together, we're talking* ***about less than a hundred*** *but they only belong to about 12 or 15 or so different species."* [6]

Dr. Luo holding a complete fossil mammal from the dinosaur era.

Perpetuating Negative Evidence

Although there is growing interest in dinosaur-era mammals, some paleontologists still focus on extracting dinosaur bones and ignore smaller mammal fossils. This practice may reflect the public's interest in dinosaurs and their lack of interest in dinosaur-era mammals.

I was surprised to learn that some dinosaur-era mammal fossils are left in the rock and not investigated or researched. It is disturbing to think that there may be a wealth of mammal bones which no one has bothered to collect, examine, investigate, or display. If this is true, it could perpetuate the illusion that mammals were rare during dinosaur times and undermine my effort to test evolution.

"We find mammals in almost all of our [dinosaur dig] *sites. These were not noticed years ago. They were very small....We have about 20,000 pounds of bentonite clay that has mammal fossils that we are trying to give away to some researcher. It's not that they are not important, it's just that you only live once and I specialized in something other than mammals. I specialized in reptiles and dinosaurs."* [7]

— Dr. Burge

Dr. Donald Burge, curator of vertebrate paleontology, College of Eastern Utah Prehistoric Museum

Mammals Found with Dinosaurs

I will now detail mammals that have been found in dinosaur layers deemed by evolution scientists to be similar in outward appearance to the modern varieties but not the same genus or species. As of this writing, I cannot confidently say that any of these mammals are modern species; but when I look at the few available skeletons, they look similar.

"Jurassic 'Beaver' From China"
American Association for the Advancement of Science
February 24, 2006

"Large Mesozoic Mammals Fed on Young Dinosaurs"
Nature
January 13, 2005

"A Cretaceous Hoofed Mammal from India"
Science
November 9, 2007

Hedgehogs and Dinosaurs

Recently, scientists discovered a mammal in Australian dinosaur layers that looked similar to a modern Southern Chinese hedgehog.[8] While I did not have access to the jaw of this modern Asian hedgehog for comparison, I did have access to the jaw of a modern European hedgehog.

If you compare the modern hedgehog jaw with the fossil, they match up fairly closely (next pages). Actually, the fossil and living hedgehog jaws are more similar than the jaws of two dogs, and we know dogs are of the same species.

"Well, as far as we can tell based on the size of the jaws and a little bit on the teeth, if you're going to take a modern animal, you'd want to pick a hedgehog in Southern China which lacks spines on it. And that's probably the closest match you can make."[8]

— Dr. Rich

Dr. Thomas Rich, curator of vertebrate paleontology at the Museum Victoria in Melbourne, Australia, holding the tiny jaw of Ausktribosphenos nyktos attached to the end of a wire.

Hedgehog-Like Mammal Found in Dinosaur Rock Layers

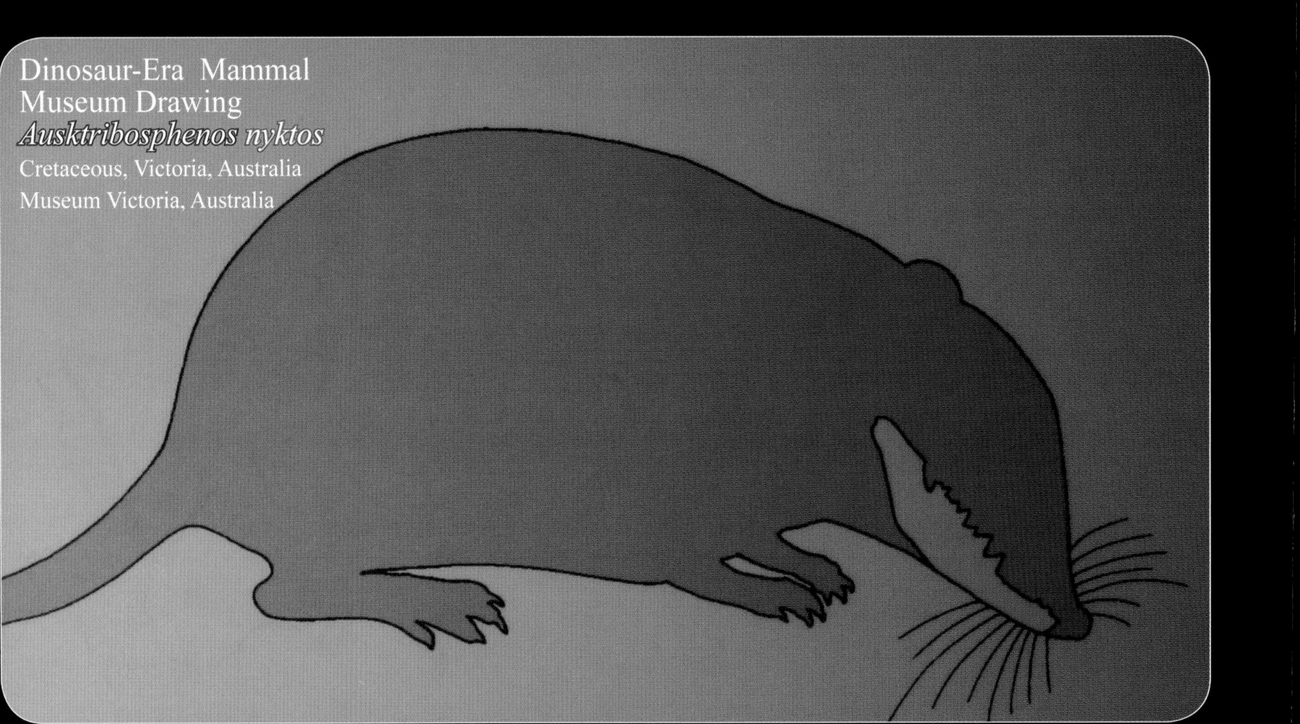

Hedgehog Jaw Similar to Dinosaur-Era Mammal Jaw

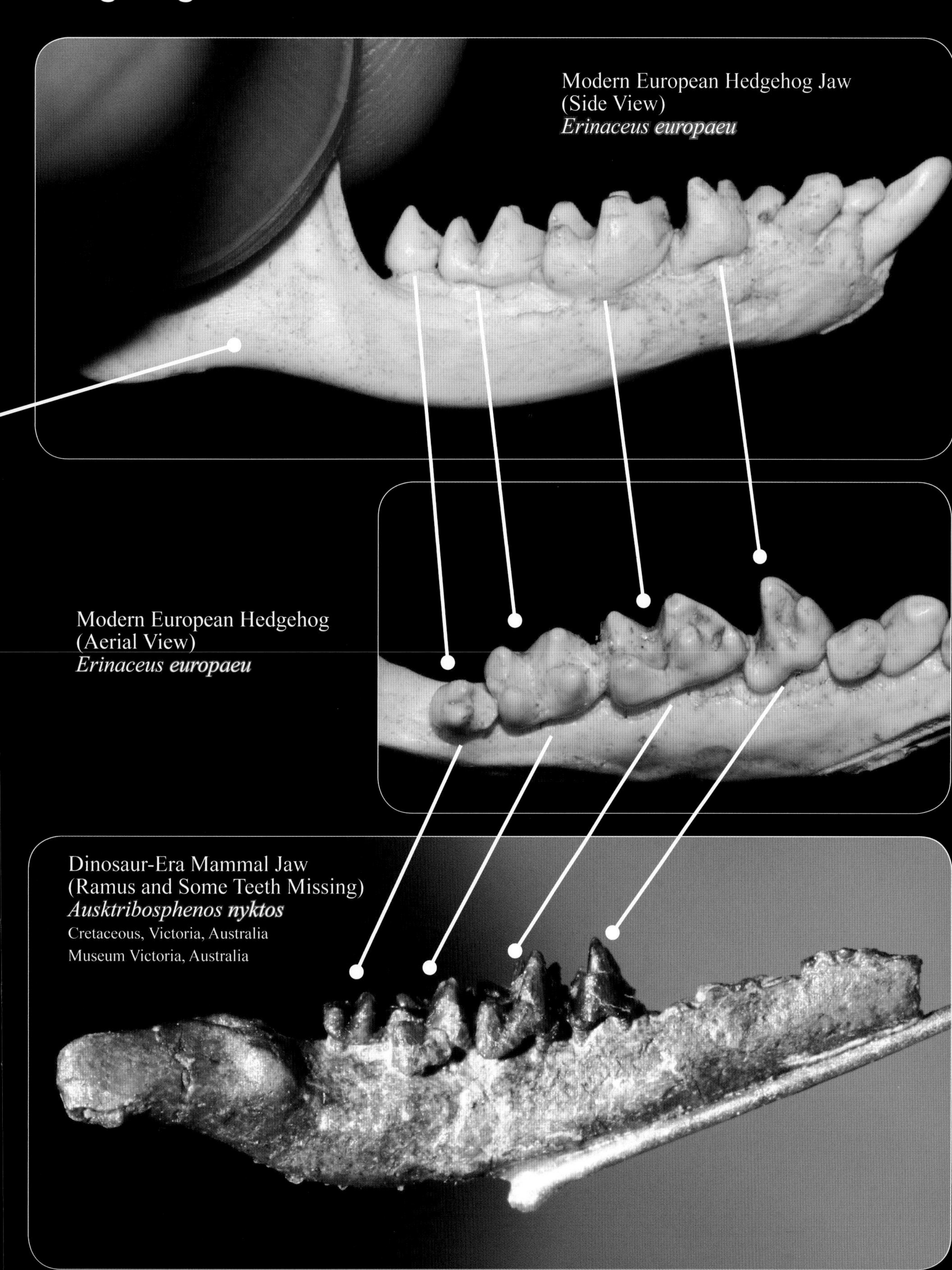

Possum-Like Mammal Found in Dinosaur Layer

Gobiconodon was found in Cretaceous dinosaur fossil layers of the Gobi Desert in Mongolia, hence the name *Gobiconodon*. Scientists who support evolution, including those at the Carnegie Museum of Natural History, have suggested that *Gobiconodon* was similar to a possum.[9, 10] At first, I discounted these claims because they seemed so preposterous — a possum living with dinosaurs — yet if true, it would lend evidence to my experiment.

I went through the work of obtaining a photograph of a ringtail possum and placed it next to the skeleton of the dinosaur-era *Gobiconodon*. For now, I will say this is a close match and have to agree with the Carnegie scientists. Look carefully at the skulls, the scapulae, the ribs, and the legs of these two specimens below. I wonder if I could make an even better match if I had a photograph of all 64 types of possums living today.

Dinosaur-Era Mammal
Gobiconodon
Carnegie Museum of Natural History, USA

See Appendix D: *Assumptions in Fossil Reconstructions*

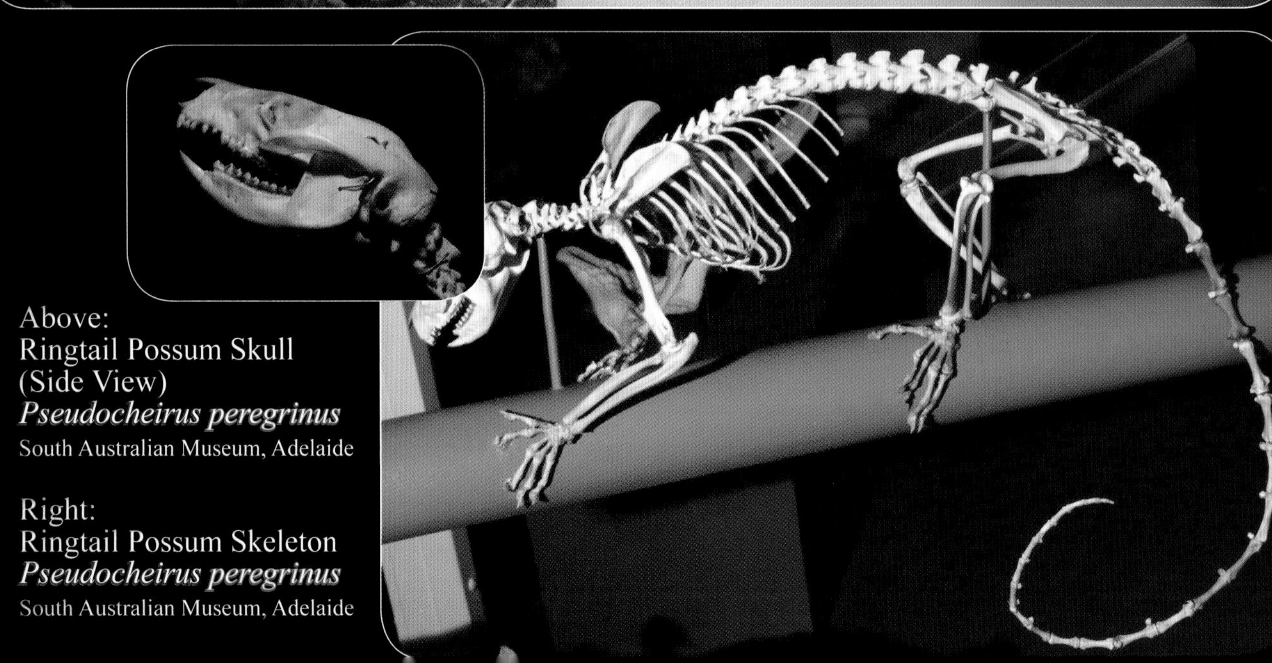

Above:
Ringtail Possum Skull (Side View)
Pseudocheirus peregrinus
South Australian Museum, Adelaide

Right:
Ringtail Possum Skeleton
Pseudocheirus peregrinus
South Australian Museum, Adelaide

Gobiconodon
California Academy of Sciences, USA

Modern Ringtail Possum
Pseudocheirus peregrinus
Australia

Three Museum Reconstructions of *Gobiconodon*

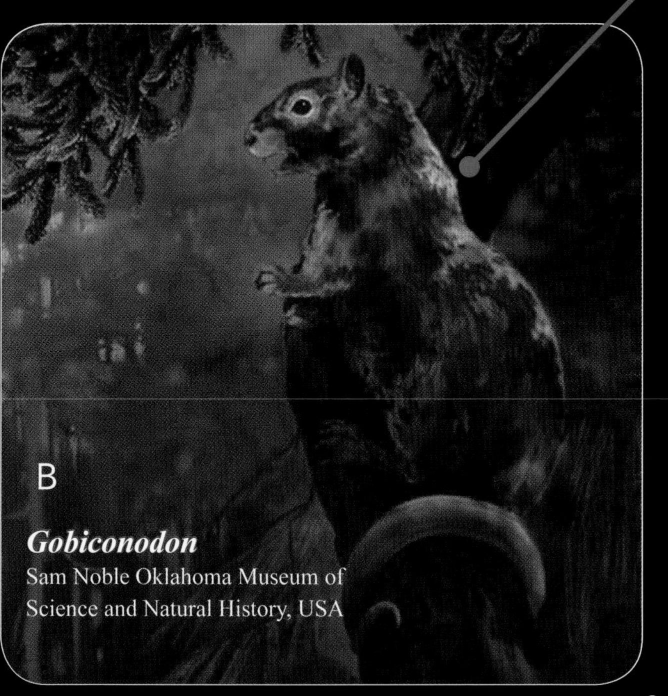

Gobiconodon
Sam Noble Oklahoma Museum of Science and Natural History, USA

Images A, B, and C are three interpretations of what *Gobiconodon* looked like at three different museums. I find it interesting that scientists could look at the same fossils and interpret their forms so differently. Images A and C make *Gobiconodon* look strange and unfamiliar, whereas Image B makes it appear similar to a modern ringtail possum (Image D). This example may help you understand the subjective nature of evolutionary ideas and natural history museum displays. With this much latitude in interpretation, a scientific illustrator could draw any dinosaur-era mammal more modern or less modern simply by changing the length of nose, the body position, the amount of fur, the length of the fur, the location of fur, the color of the fur, and the size and shape of the ears.

See Appendix D: *Assumptions in Fossil Reconstructions.*

Gobiconodon
Carnegie Museum of Natural History, USA

Duck-Billed Platypus, Echidnas, and Shrews

"*The opossums of our woods would have been at home in Mesozoic* [dinosaur] *times. Likewise, the duck-billed platypus and echidna, now of the Australian region, would not have been out of place. Even our shrews, small insectivorous mammals, had Mesozoic counterparts.*"[11]

— Dr. Dawson

Dr. Mary Dawson, paleontologist,
Carnegie Museum of Natural History

Virginia Opossum
Didelphis virginiana
South Carolina, USA

Modern Tree Shrew
Tupaia tana

Modern Echidna (Spiny Anteater)
Tachyglossus aculeatus
Kangaroo Island, Australia

Modern Duck-Billed Platypus
Ornithorhynchus anatinus
Milwaukee Public Museum, USA

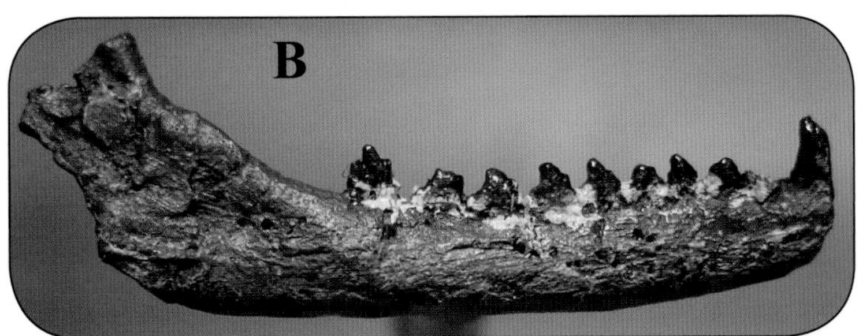

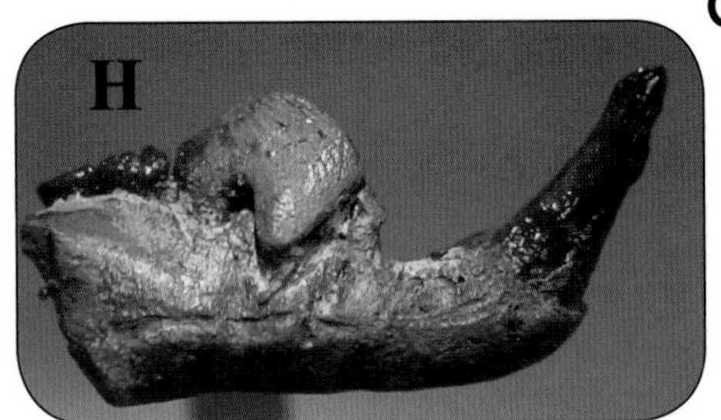

Other Mammals Found in Dinosaur Layers

Clockwise from top left:
A: A small Jurassic mammal, a Triconodont. **B:** *Docodon*, a Jurassic mammal. **C:** *Sinosdelphys*, a Cretaceous marsupial mammal with teeth similar to a modern opossum. **D:** *Amblotherium*, a Jurassic mammal. **E:** *Asioryctes*, a Cretaceous placental mammal. **F:** Jaw of *Didelphodon,* a Cretaceous marsupial. **G:** *Priacodon*, a Jurassic mammal. **H:** *Cimolodon*, a Cretaceous mammal.

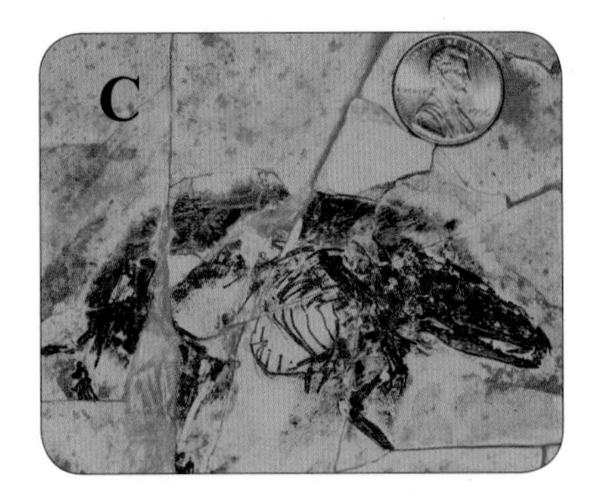

G

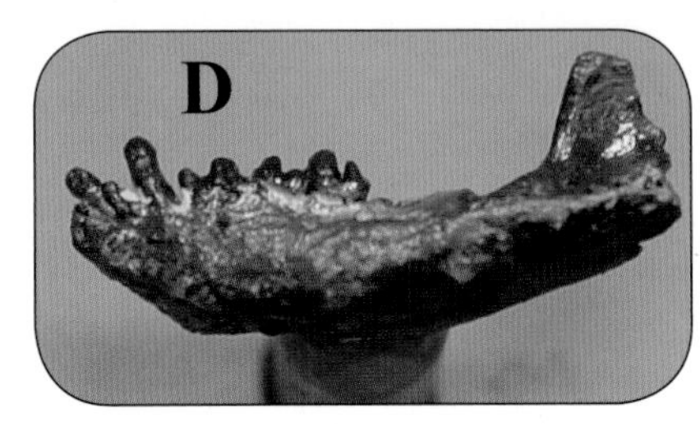

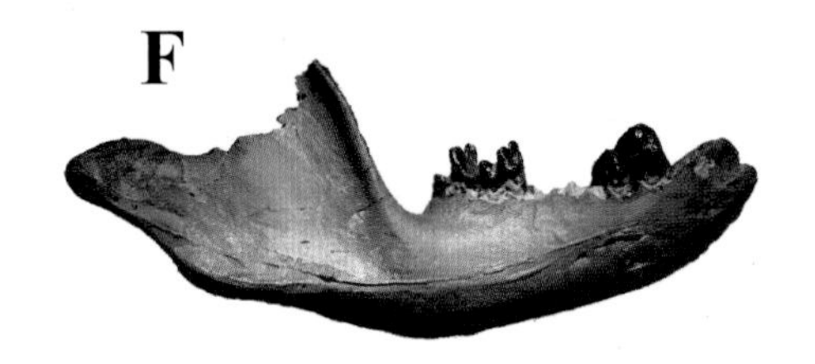

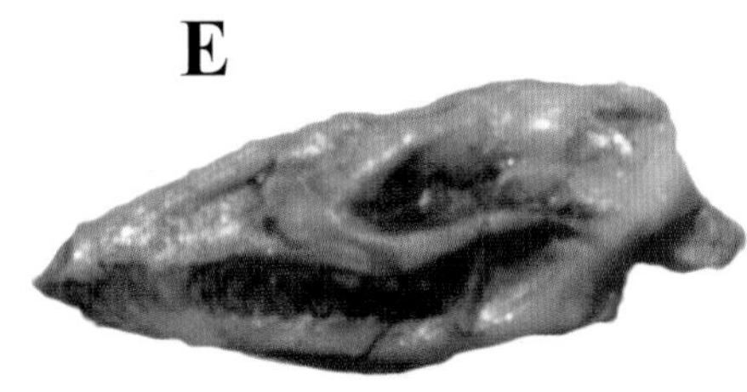

Summary

The animals presented in this chapter represent all three groups of mammals living today, namely the marsupials, the placentals, and the egg-laying montremes.[12, 13] I have presented the fossils of two *modern-appearing* mammals, a ringtail opossum-like animal and a hedgehog-like mammal. Besides these there were others such as a "Jurassic 'Beaver' from China,"[14, 15] a dinosaur-eating Tasmanian Devil-like *Repenomamus*,[16] a duck-billed platypus-like mammal, an echidna-like mammal, a shrew-like mammal and various small insectivore mammals. I cannot verify if any of these other dinosaur-era mammals were truly similar to modern mammals since I was unable to photograph these fossils. I never saw them on display.

Cone-Bearing Plants

This chapter deals with the three groups of cone-bearing plants living today, namely cone trees (Division Coniferae), palm-like cycads (Division Cycadophyta), and gingko trees (Division Ginkgophyta).

Chapter 22

"After the interview, Debbie was ready to see Mount Rushmore, but I had other ideas."

How Mount Rushmore Almost Cost Me the Challenge

If you asked most people whether modern plants were alive during the time of the dinosaurs, they would say no. This is what I believed for decades. ***My experiment predicted the opposite — that I should find modern plants with dinosaurs if evolution was not true.***

While few would list the Museum of Geology in Rapid City, South Dakota in the top 50 museums of the world, it helped me carry out my experiment regarding plants. Up until this point, I was lacking fossil evidence concerning plants and this part of my experiment seemed destined for failure. None of the previous museums we visited had displayed dinosaur-age fossil plants to any meaningful degree.

The cabinet containing most of the dinosaur-era plant fossils at this museum (shown in the photo below) is probably only 4 feet wide by 15 feet long and not marked with a large placard referencing how important these finds are. In fact, the cabinet is so modest and insignificant-looking, I almost missed it.

I spotted this case late in the day, after an exhausting interview, just as we were getting ready to leave the museum and go sightseeing. Debbie was ready to see Mount Rushmore, just a few miles away, but once I saw this cabinet, I couldn't pass it up.

Below: *The one-room Museum of Geology in Rapid City, South Dakota contained a treasure trove of dinosaur-era fossil plants in the small display case on the left.*

Above: *Debbie and I went to visit Mount Rushmore after photographing fossil plants at the Museum of Geology in Rapid City.*

I talked Debbie into going back to the RV on the museum parking lot and resting there with our Golden Retriever, Curly, while I photographed these fossil plants. She agreed. But as the *hours* clicked by, Debbie became more and more impatient. Finally, she called me on my cell phone and asked, "*What are you doing? You have been photographing the fossils in that one tiny case for two-and-a-half hours!*" I explained to her that I was taking special care to photograph each of these fossil plants. She seemed exasperated, as I am sure anyone would be. I was *again* beginning to feel a little more than foolish over this whole endeavor. (I sometimes wonder what my evidence for plants would have been had I not persisted.)

The museum cabinet had a variety of dinosaur-era fossil plants, which are presented in the next three chapters of this book. This chapter deals with cone-bearing plants, also called gymnosperms.

Did Modern Cone Trees Exist at the Time of the Dinosaurs?

Redwoods

Coast Redwood
Sequoia sempervirens
Muir Woods National Monument
California, USA

Sequoia Trees Found in Cretaceous Dinosaur Layers

The redwood trees (sequoias) in California are ectacular, growing up to 300 feet tall. Few would vision dinosaurs roaming among them. The fact is, quoias were living at the time of the dinosaurs.

There are four live redwood cones at the bottom ht corner of this page and one *fossil* cone below. ı the live cones, you can see oval seed scales (X) ojecting outward and spaces (0) between the seed ales.

The fossil cone appears to be a *negative cast* of the live cone. In a cast, everything is in reverse. The oval seed scales (X) are absent, and the spaces between the seed scales (0) are filled in with sediment.

In the four live cones, notice how much they differ in height and width. The fossil seems to fall within the range of variability as seen in the live cones. Now compare the species names.

nosaur-Era Sequoia Cone
egative Cast of Cone
quoia langsdorfii
etaceous, Montana, USA
useum of Geology - South Dakota School
Mines and Technology, Rapid City, USA

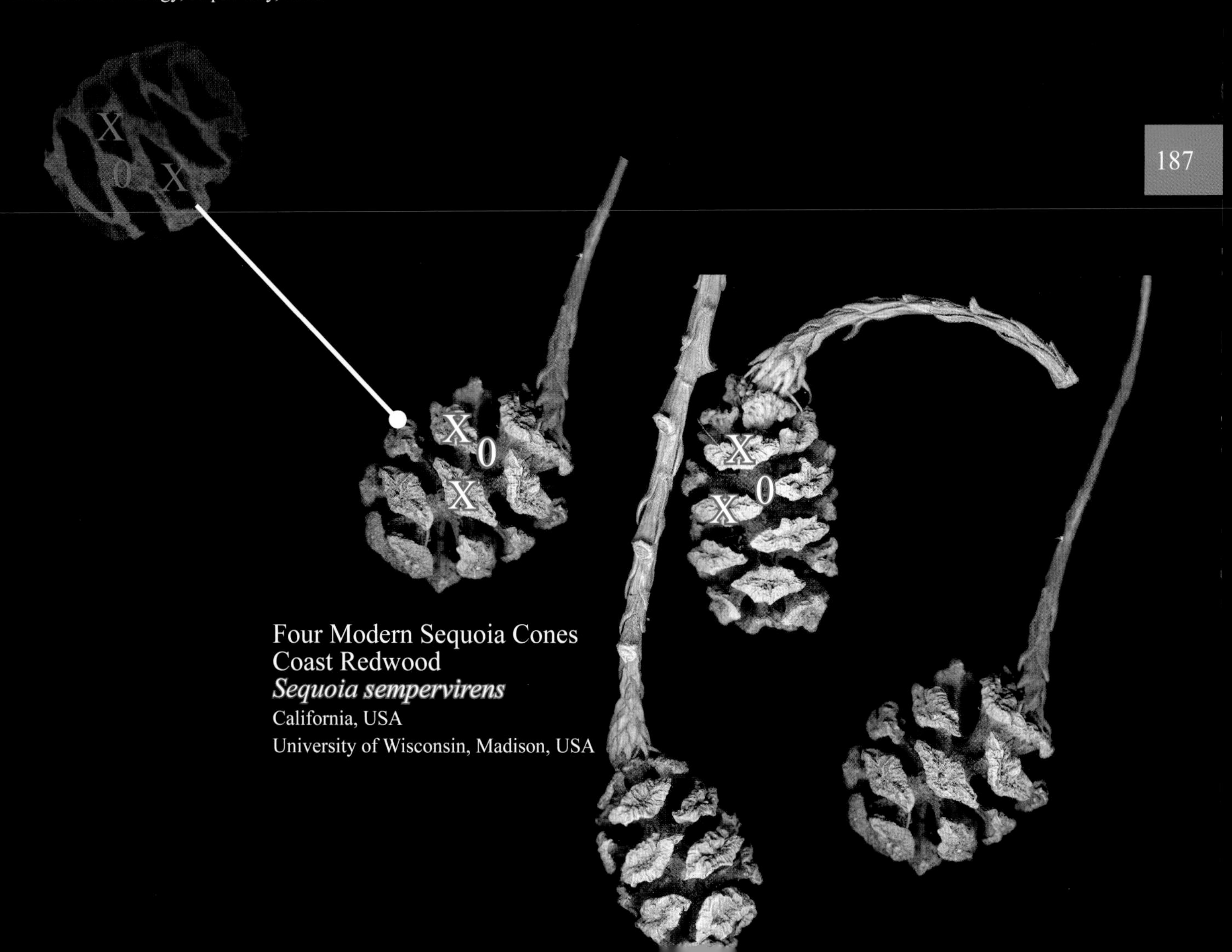

Four Modern Sequoia Cones
Coast Redwood
Sequoia sempervirens
California, USA
University of Wisconsin, Madison, USA

Dawn Redwoods

Dawn Redwoods
Metasequoia glyptostroboides
Missouri Botanical Gardens, USA

Dawn Redwood Trees Found at Dinosaur Dig Site

For years, scientists considered the Dawn Redwood to be an extinct fossil tree. Then, in 1941, a Chinese forester discovered a *live* Dawn Redwood tree. When scientists compared the living tree to the fossils (below), they realized it was "almost identical." [1] Notice again the variety in the details of the four cones of the single modern species.

Dawn Redwood
Branch and Four Cones
Metasequoia glyptostroboides
Missouri Botanical Gardens, USA

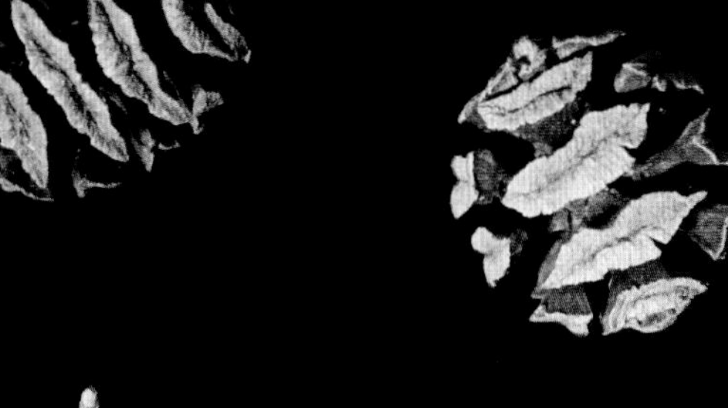

Dinosaur-Era Dawn Redwood
Fossil Branch and Cone
Metasequoia cuneata
Cretaceous, Hell Creek, Montana
Milwaukee Public Museum, USA

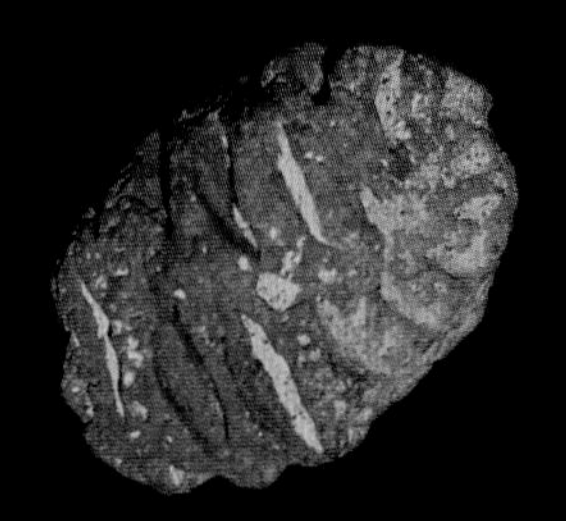

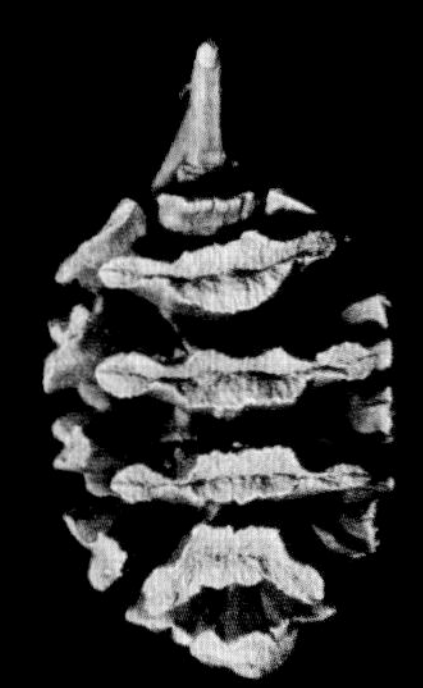

Cook Pine Cones

When Debbie photographed the beautiful fossil pine cones on the next page, I remember thinking I had never seen anything like this before. Believing the fossil cones to be an extinct species, I doubted I would ever find a modern match — but I persisted.

I presented photographs of the fossil cones to a botanist at the Missouri Botanical Gardens and asked if he had ever seen a similar, but living, type of pine cone. He gladly took my photograph and said he would get back to me.

It wasn't long afterwards that I was beckoned back to the gardens. This kind botanist proceeded to take me to a basement room where scores of cabinets filled with dried pine cones from around the world were assembled. It was quite an impressive collection. He walked to one cabinet, opened a drawer, and pulled out the two modern Cook Pine cones on this page, collected in New Caledonia, and placed them next to the photos. We both smiled.

Without saying anything more, look at the species names (in red) of the fossil and living pine cones.

Modern Cook Pine Cones
Araucaria columnaris
Collected in New Caledonia,
Missouri Botanical Gardens, USA

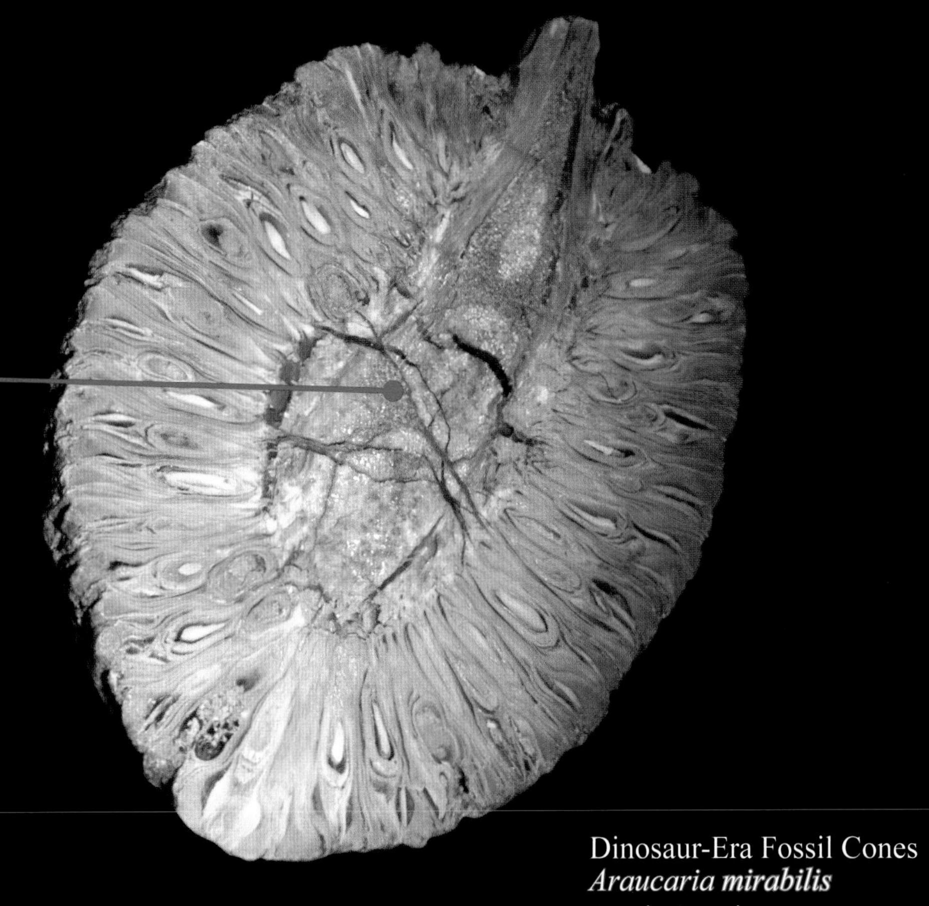

Dinosaur-Era Fossil Cones
Araucaria mirabilis
Jurassic, Argentina
Geological Museum,
University of Wisconsin, Madison, USA

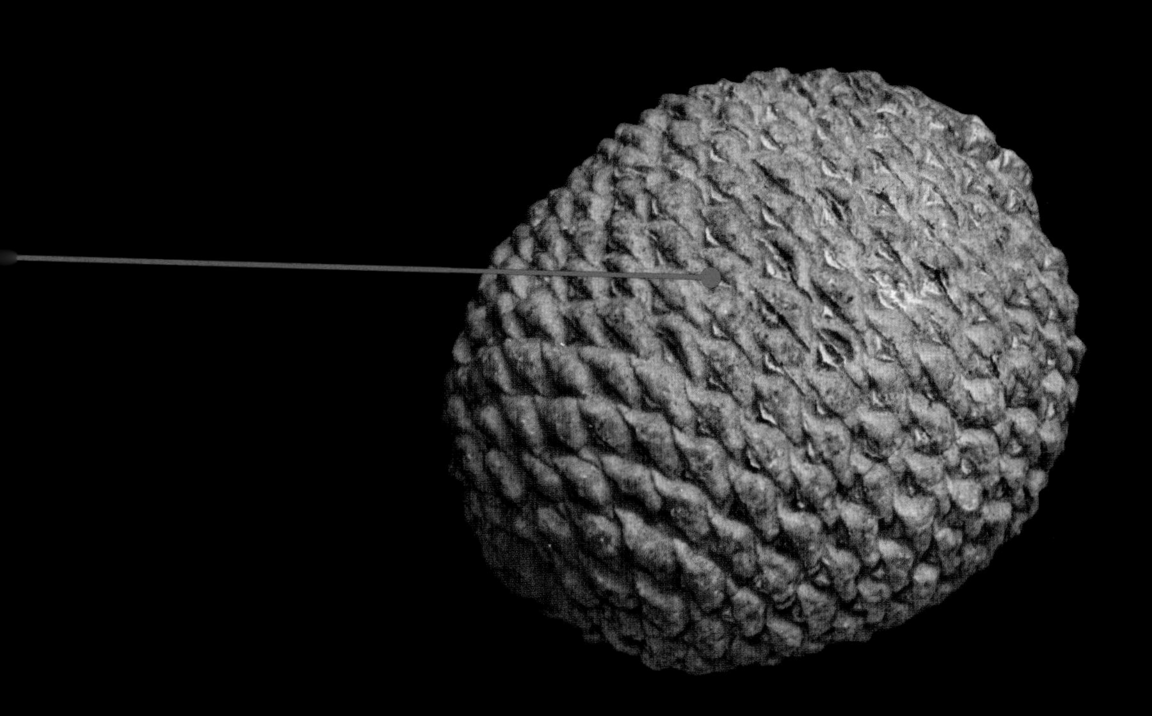

Bald Cypress Trees

When I was a kid, my dad and I went fishing at Reelfoot Lake in Tennessee. We loved to fish with crickets around the "knees" (roots) of bald cypress trees that lined the shore of the lake. Sometimes we would see ducks and gar in the shallows around the trees. I had no idea then, but now I realize that all of these creatures — ducks, crickets, gars and bald cypress trees — also lived with the dinosaurs. (See museum display on Page 227.)

A modern bald cypress knee (root) sticking up out of water and a modern duck (far left).

Modern Swamp
Bald Cyptress Tree and Duck
Mississippi Museum of Natural Science, USA

Pine Tree

This tree branch with needles is an Eastern White Pine tree. Compare this branch to the fossil below. I have to conclude that modern-appearing trees with needles lived during the time of the dinosaurs.

Modern White Pine
Pinus strobus
Missouri, USA

Dinosaur-Era Conifer
Unidentified species
Cretaceous, California, USA
Museum of Paleontology,
University of California, Berkeley, USA

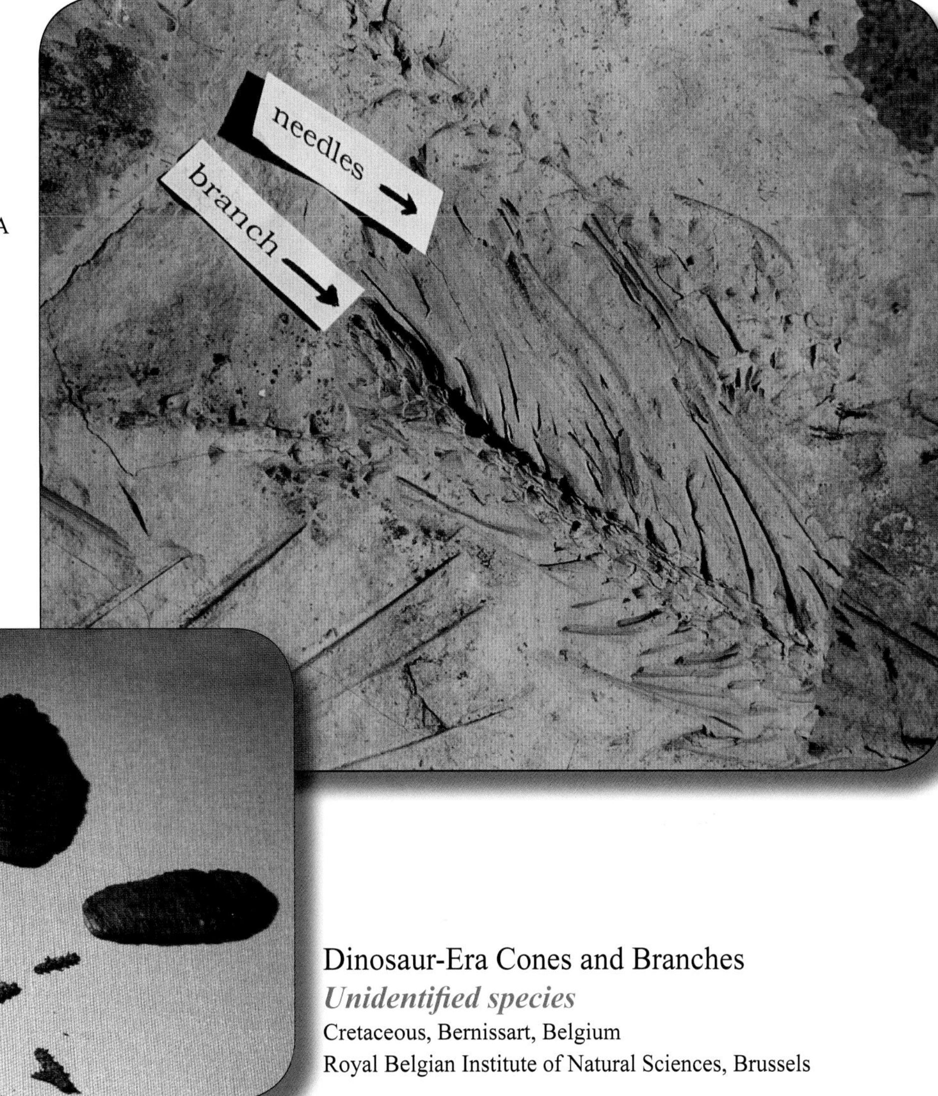

Dinosaur-Era Cones and Branches
Unidentified species
Cretaceous, Bernissart, Belgium
Royal Belgian Institute of Natural Sciences, Brussels

Cycads

Cycads are palm-like cone-bearing plants that grow in tropical and subtropical areas. Debbie's parents have one growing in their front yard in Louisiana, similar to the one on the title page of this chapter. Compare this living cycad trunk (right) to a fossil cycad trunk (below) and notice their genus and species names.

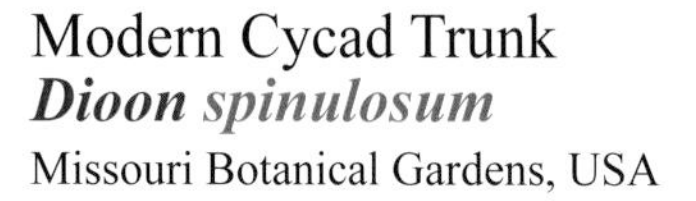

Modern Cycad Trunk
Dioon spinulosum
Missouri Botanical Gardens, USA

Dinosaur-Era Cycad Trunk
Cycadeoidea (species unidentified)
Cretaceous, Cheyenne River,
Museum of Geology - South Dakota
School of Mines and Technology, USA

Coontie
***Zamia** pumila*
Missouri Botanical Gardens, USA

Coontie

The Coontie is a small, woody cycad native to Cuba and the Dominican Republic. Compare the modern Coontie branch above to this fossil cycad branch found in Jurassic fossil layers in France (right).

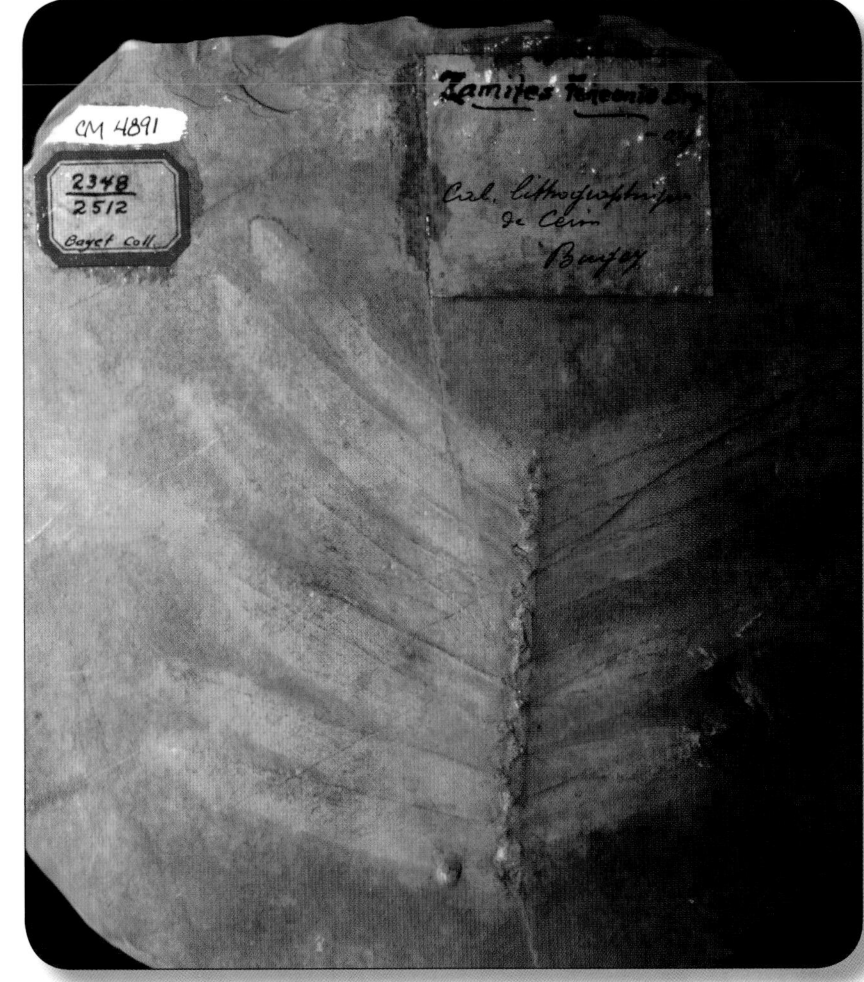

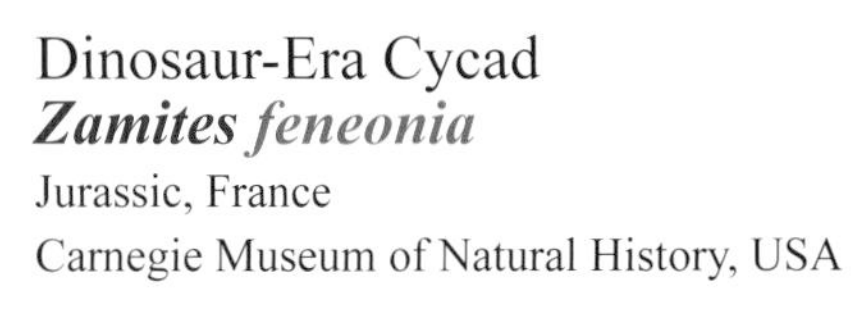

Dinosaur-Era Cycad
***Zamites** feneonia*
Jurassic, France
Carnegie Museum of Natural History, USA

Modern Ginkgo with Solid Leaves
Ginkgo biloba
Missouri Botanical Gardens, USA

Dinosaur-Era Ginkgo with Solid Leaves
Ginkgo adiantoides
Cretaceous, Montana, USA
Milwaukee Public Museum, USA[2]

Maidenhair Tree (Ginkgo)

Ginkgos are the third major type of cone-bearing plants. As seen in the two photos on the left side of this page, there are great variations in the splitting of leaves in this one species of Ginkgo living today, *Ginkgo biloba*. Some leaves have no splits and look like a fan (top, left) ; others have two or more splits (bottom, left). Compare these living *Ginkgo biloba* leaves on the left side of this page to these dinosaur-era Ginkgo leaves on the right side of this page. Now, compare their names.

Modern Ginkgo with Split Leaves
Ginkgo biloba
Missouri Botanical Gardens, USA

Dinosaur-Era Ginkgo with Split Leaves
Ginkgo digitata
Cretaceous, Norway
Palaeontological Museum, University of Oslo, Norway

Summary

Examples of all three divisions of cone-bearing plants living today, conifers, cycads, and Ginkgos, were also alive during the time of the dinosaurs.

Spore-Forming Plants

This chapter deals with plants that reproduce with spores, such as ferns and horsetails (Division Pteridophyta), moss (Division Bryophyta), and club moss (Division Lycopodiophyta).

Chapter 23

"We photographed a fossil fern in a museum, and then, a few days later, we spotted it in the rainforest."

Ferns in the Australian Rainforest

I had always read that ferns were "prehistoric plants" and that the coal beds mined in the United States are heavily laden with ferns. But I mused, *"If they are 'prehistoric', why are they still around today? And why were they not eliminated according to the principle of the survival of the fittest?"* Even more importantly, if evolution is true, why do ferns appear essentially unchanged, contradicting the evolutionary idea of major change over time? Let me show you an example.

In 2006, Debbie and I traveled to Australia to conduct some interviews in Melbourne and Adelaide and also to photograph and dive at the Great Barrier Reef near Cairns. After filming at some of the museums along the southern coast of Australia, we traveled northwest to Cairns where we went on an eco-tour through the tropical rainforest. (Yes, there is a rainforest in Australia very close to Cairns.) Our guide (below), an amusing Aussie who resembled Crocodile Dundee (with his hat, personality, and all), drove us through the rainforest on an amphibious duck and pointed out the various plants along the way.

It was a strange moment when I spotted a fern that reminded me of the one we had just photographed in Melbourne (next page). As you can see for yourself, they look nearly the same.

Black Trunk Tree Fern

Dinosaur-Era Fern
Cladophlebis australis
Cretaceous, Victoria, Australia
Museum Victoria, Melbourne, Australia [1]

Black Trunk Tree Fern
Cyathea medullaris
Australian Rain Forest
Cairns, Australia [2]

Giant Tree Fern

Cyathea cooperi
Mainland Australia, Rainforest
Cairns, Australia

Tree Ferns Found in Australian Dinosaur Layers

Minutes later, the guide drove us past a tree fern (previous page). There are a variety of tree ferns living today on the mainland of Australia and on the Australian island of Tasmania. Look at this fossil below and compare it to this living tree fern branch from Tasmania. They are similar. Notice the species names and also the genus names.

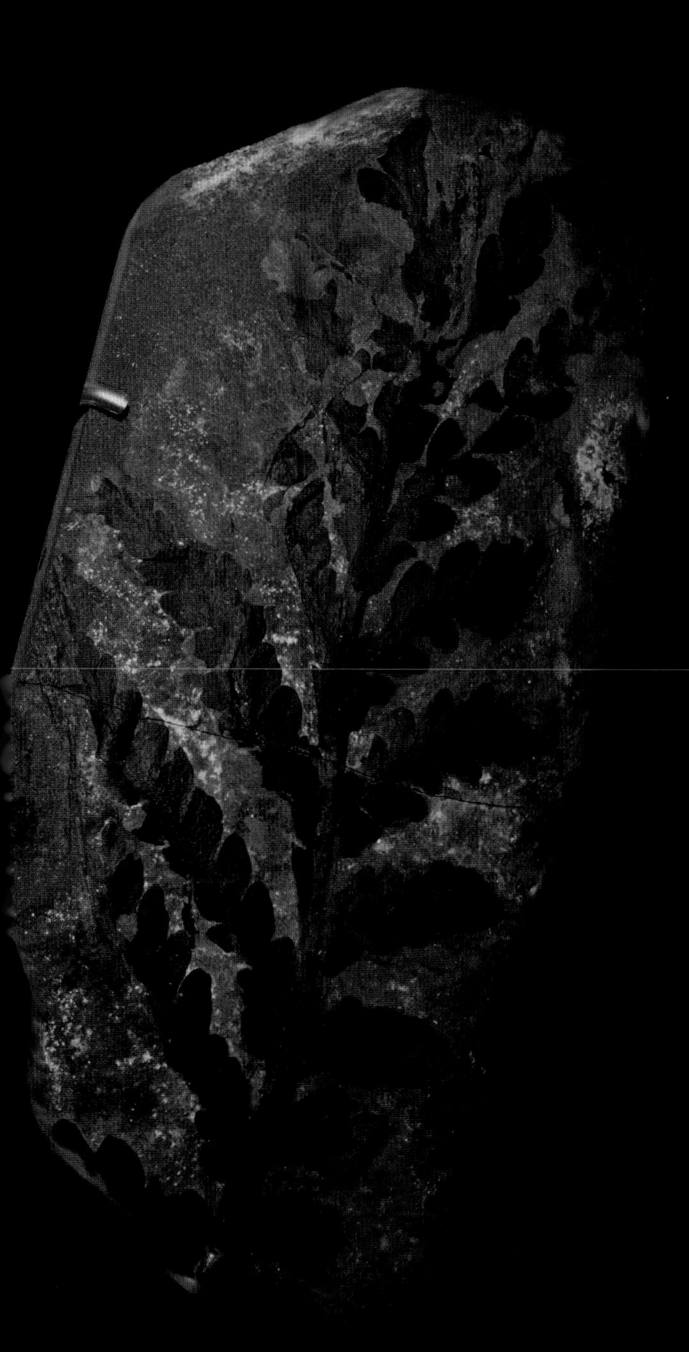

Dinosaur-Era Fern
Thinnfeldia feistmanteli
Triassic, Location Unidentified
Museum Victoria, Melbourne, Australia

Tasmanian Tree Fern
Dicksonia antarctica
Collected in Tasmania, Australia
Missouri Botanical Gardens, USA

Sensitive Ferns

At the Public Museum of Natural History in Milwaukee, there is a display of a sensitive fern that was found with a dinosaur in Montana. Compare this dinosaur-era plant model (below, left) to the living sensitive fern (below, right). Frankly, I can't tell them apart. Once again, notice the genus names in blue.

Dinosaur-Era Sensitive Fern
Museum Reconstruction
***Hydropteris* pinnata**
Cretaceous, Montana, USA
Milwaukee Public Museum, USA[4]

Living Sensitive Fern
***Onoclea* sensibilis**
Missouri Botanical Gardens, USA

Other Ferns

Modern Wood Fern
***Dryopteris** lacera*
Missouri Botanical Gardens, USA

Dinosaur-Era Fern
***Phlebopteris** smithii*
Triassic, Arizona, USA
Petrified Forest National Park, USA

Eoraptor lunensis
Triassic, Argentina
Missouri Botanical Gardens, USA
Lost World Studios, Missouri, USA

Mosquito Fern
***Azolla** caroliniana*
Missouri Botanical Gardens, USA

Right: *Mosquito ferns such as these have been found as fossils in dinosaur layers and are "almost identical" to modern forms.*[5]

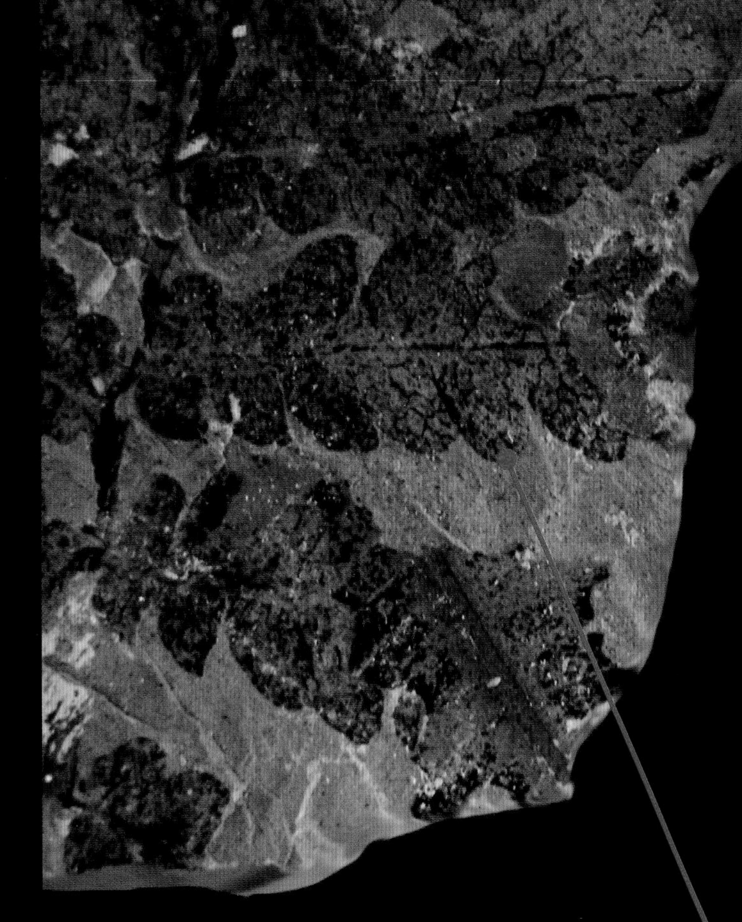

Dinosaur-Era Fern
Cynepteris lasiophora
Triassic, Arizona. USA
Petrified Forest National Park, USA [6]

Shield Fern

Shield Fern
Polystichum bicknellii

Horsetails Found at Dinosaur National Monument

Horsetails (scouring rush) have a hollow, jointed stem. They live in moist environments and reproduce by spores. Look at this fossil horsetail found at Dinosaur National Monument and compare it to the living horsetail.

Dinosaur-Era Horsetail
Unidentified species
Jurassic, Utah, USA
Dinosaur National Monument, USA

Modern Horsetail
Unidentified species
Missouri, USA

Club mosses are unique because they have a vascular system like a tree yet they reproduce with spores. Look at this "extinct" fossil club moss (below, left) found in the pre-dinosaur fossil layers and compare it to the modern club moss (below, right). Now compare the genus names.

Fossil Club Moss
Lepidodendron obovatum
Carboniferous, Austria
Geological Museum, University of Wisconsin, Madison, USA

Club moss
Lycopodium selago
Scotland

Above: *These dinosaurs, on display at the Houston Museum of Natural Science, lived at the same time as club moss, peat moss, and ferns. Despite the passing years, these plants are similar in appearance to the dinosaur-aged fossil varieties.*

Peat Moss

Sphagnum moss, also known as peat moss, does not have a vascular system like club moss. Instead, it absorbs moisture from its surroundings. The raised bogs in Ireland today are "almost pure sphagnum moss...and these are the bogs from which moss peat is extracted." [7]

Sphagnum moss was found fossilized in dinosaur rock layers (see Page 228). Imagine this: Peat moss that we use in our gardens is similar to the moss dinosaurs walked on. It's beginning to feel like the movie "Jurassic Park."

Peat Moss
Sphagnum palustre
Missouri Botanical Gardens, USA

Summary

My experiment predicted that I should find modern types of spore plants in dinosaur rock layers if evolution was not true. I found fossil evidence of all three types of spore-producing plants, ferns and horsetails, moss, and club moss. Now onto the last and most difficult test of all, flowering plants.

Flowering Plants

This chapter deals with flowering plants (Division Magnoliophyta).

Chapter 24

"After watching Carl Sagan's 'Cosmos', I thought I lost the challenge."

Was Carl Sagan Wrong? Or Is Carl Werner Wrong? You Decide!

According to the theory of evolution, flowering plants (also called angiosperms) were the most sophisticated of all plants to have ever lived on the earth and the last type of plants to evolve. In evolutionary terms, they would be considered newcomers on the block.

Most plants that you are familiar with are flowering plants. Examples include sassafras trees, oak trees, palm trees, lilies, lily pads, grapes, rhododendrons, tomatoes, etc.

One year after taking on the challenge, Carl Sagan host of the blockbuster television series "*Cosmos*," told his worldwide audience of 500 million viewers: "*The dinosaurs perished around the time of the first flower.*" [1] When I heard this, it made me think that evolution was, in fact, true. I reasoned that if dinosaurs perished at the time that flowering plants were just getting started, then dinosaurs would have never seen modern types of flowering plants. If plant life was completely different during the time of the dinosaurs (as compared to today) then plants changed dramatically over time. And if this was true, then the evolution of plants from one type into another occurred. Can you think of any other way to interpret Sagan's statement? For me, this was a simple but elegant evidence for evolution.

"The dinosaurs perished around the time of the first flower." [9]

— Dr. Sagan

Above: *Museum display at the Oklahoma Museum of Science and Natural History.*

College textbooks that I subsequently read made similar statements. In the 1996 edition of ***The Earth through Time***, the author wrote, "*Although a vertebrate paleontologist might refer to the* [dinosaur era]...*as the Age of Reptiles, paleobotanists might well argue that the term 'Age of Cycads' would be equally appropriate. The cycads...are seed plants in which true flowers have not been developed.*"[2] This author's statement again implied that dinosaurs lived with the non-flowering cycad plants.

Museums also project this idea. Look at this display from the Oklahoma Museum of Science and Natural History (above). It implies to me that flowering oak trees and sassafras trees (angiosperms) did not live until after the Cretaceous — the last dinosaur layer.)

Museum displays, textbooks, and world-class scientists, such as Carl Sagan, insinuated to me that my experiment was over. But when I reached the Museum of Geology in Rapid City, South Dakota, I realized something was wrong.

What I am about to show you is that modern types of flowering plants and trees were alive during the time of the dinosaurs, and *lots* of them. The three previous statements were misleading at best... or just plain wrong.

Fossil Flowers From Dinosaur Rock Layers

Look at these perfectly preserved fossil flowers from dinosaur rock layers on display at two different museums. Few would appreciate their significance in terms of evolutionary history. But for me, these fossilized flowers contradicted Sagan's statement.

Dinosaur-Era Flower
Cretaceous, Nebraska, USA
University of Nebraska State Museum, Lincoln, USA

Dinosaur-Era Flowers Preserved in Charcoal
Cretaceous, Unidentified Location
Sam Noble Oklahoma Museum of Science
and Natural History, USA

Flowering Plants Found in Dinosaur Layers: Rhododendrons

The first modern type of flowering plant I want to show you is on display (below) at the Carnegie Museum. Scientists at this museum reconstructed this flowering plant based on fossils found in dinosaur rock layers. When I saw this display, I immediately thought, "*What? Rhododendrons living with dinosaurs? Debbie and I have these growing in our backyard.*" So began my list of common flowering plants found in dinosaur layers that seemingly repudiate the idea that the dinosaur times were the "age of the *non-flowering* cycads." (Other dinosaur-era dig sites also have recovered rhododendron fossils.[3])

Dinosaur-Era Rhododendron
Museum Reconstruction Based on Fossils
Cretaceous, Unidentified Location
Carnegie Museum of Natural History, USA

Poppies

Flowering poppies, similar to those living today, have also been found in dinosaur layers and are currently on display at the Carnegie Museum. [4]

Lily Pads

Compare this museum reconstruction of a fossilized lily pad (or lotus) to a modern flowering lotus. Remember, color should always be ignored in fossil reconstructions.*

Dinosaur-Era Lotus
Museum Reconstruction
Nelumbium montanum
Cretaceous, Montana, USA
Milwaukee Public Museum, USA[5]

Modern Lotus
Missouri, USA

***See Appendix D:** *Assumptions in Fossil Reconstructions — Color.*

Sweetgum

The sweetgum tree is a beautifully shaped flowering tree that grows in the Midwest. My father had one in his yard. Each fall this pesky tree drops hundreds of prickly gumballs — a homeowner's nemesis.

Sweetgum
Liquidambar styrafciflua
Missouri, USA

Dinosaur-Era Leaf
Unidentified species
Cretaceous, Kansas
Geological Museum, University of Wisconsin, Madison, USA

Sassafras

Remember the tremendous variations in sassafras leaves on a single tree as seen on Page 25? Now compare this fossil leaf found in Cretaceous dinosaur layers in Kansas to the living sassafras leaf below. Notice the species names in red.

Dinosaur-Era Sassafras Leaf
Sassafras mudgei
Cretaceous, Kansas, USA
Museum of Geology - South Dakota School of Mines and Technology, Rapid City, USA

Sassafras Leaf
Sassafras albidum
Missouri, USA

Poplar

Poplar trees are tall, narrow, flowering trees that are sometimes used to mark property lines. Compare the modern poplar leaf to the fossil form and then compare the species names in red.

Living Poplar Leaf
Populus szechuanica
France[6]

Dinosaur-Era Poplar Leaf
Populus kansaseana
Cretaceous, Kansas, USA
Museum of Geology - South Dakota School of Mines and Technology, Rapid City, USA

Walnut

Black Walnut
Juglans nigra
Missouri, USA

Dinosaur-Era Leaf
Dryophyllum subflactum
Cretaceous, Montana, USA
Museum of Geology - South Dakota School of Mines and Technology, Rapid City, USA

Ash Leaf
Fraxinus excelsior
France[7]
Dinosaur-Era Ash Leaf
Fraxinus leei
Cretaceous, South Dakota, USA
Museum of Geology - South Dakota School
of Mines and Technology, Rapid City, USA
Ash Tree

Modern Chinese Soapberries
Sapindus mukorossi

Soapberry

The fruit of soapberries are bright red and feel soapy to the touch. They are used today to make natural soap. Leaves from this flowering plant were found in dinosaur rock layers.

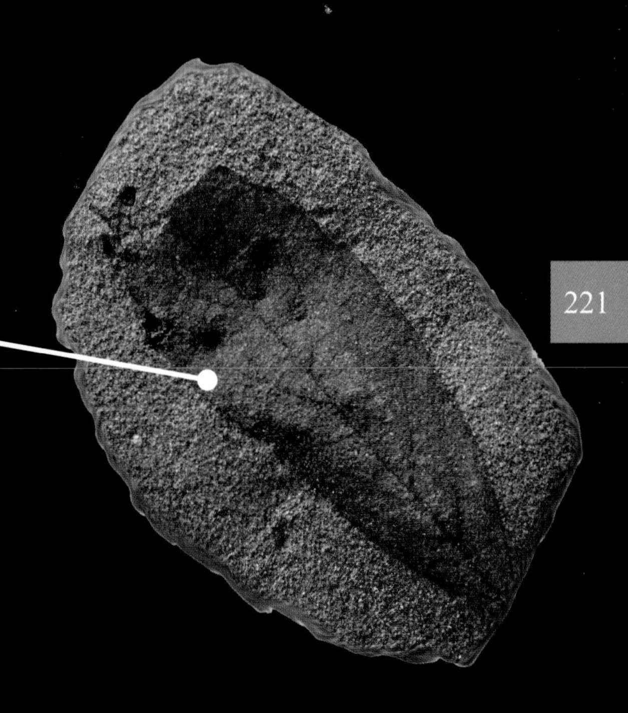

Dinosaur-Era Leaf
Sapindus (species unidentified)
Cretaceous, South Dakota, USA
Museum of Geology - South Dakota School of Mines and Technology, Rapid City, USA

Modern Soapberry Branch
Sapindus saponaria
Photo courtesy Department of Forestry Virginia Tech[8]

Bay Leaves

Bay leaves were part of the wreaths worn by the poets and emperors of Greece and Rome. The terms poet *laure*ate and bacca*laure*ate refer to these *laurel* (bay) leaves. Apparently, the dinosaur-era bay leaves look similar to those worn by Greek emperors.

Dinosaur-Era Bay Leaf
Unidentified species
Cretaceous, Unidentified Location
University of Nebraska State Museum, Lincoln, USA

Modern California Bay Leaf
Umbellularia californica
California, USA[9]

Viburnum Leaf
Viburnum carlesii
Missouri Botanical Gardens, USA

Above: *Viburnum Snowball Bush*

Viburnum

One popular type of viburnum is the "snowball" bush. My parents had one in their backyard. I see little difference between this dinosaur-era viburnum and modern varieties.

Dinosaur-Era Viburnum Leaf
Viburnum longifolium
Cretaceous, Central Kansas, USA
Museum of Geology - South Dakota School of Mines and Technology, Rapid City, USA

"I always think about the late Cretaceous [late dinosaur era], *that if you were there and were able to wander around in those forests, the vegetation would have looked really very much like today: Dogwoods, tulip trees, magnolias and so on, walnuts, oak relatives, chestnut relatives and so on, all of these groups that we would think of as being fairly modern groups of plants also being around quite a long time ago."* [10]

—Dr. Crane

Dr. Peter Crane, curator, Royal Botanic Gardens, London, England

Oak and Dogwood

According to Dr. Peter Crane, oak trees were present in Cretaceous dinosaur layers. This seems to contradict the museum display on Page 211 and Dr. Sagan's statement.

Oak

Dogwood

Dinosaur-Era Magnolia
Magnolia actifolia
Cretaceous, New Mexico, USA
New Mexico Museum of Natural History and Science, USA

Magnolia
Magnolia yunnanensis
JC Raulston Arboretum
North Carolina State University, USA[11]

Magnolia and Chestnut

Chestnut

Moreton Bay Fig,
Australian Banyan
Ficus macrophylla

Museum Diagram Showing Other Types of Modern Trees in Cretaceous Forests

Screw Pine
Pandanus
Missouri Botanical Gardens, USA

Dinosaur
Ornithomimus
Cretaceous
Utah Field House of Natural History, USA

Everglades Palm
Paurotis wrightii
Missouri Botanical Gardens, USA

West Indian
Thatch Palm
Thrinax radiata
Florida, USA[12]

Indian Laurel
Ficus nitida
California, USA[14]

Madagascar Palm
Pachypodium lamerii
Russia[15]

Above: *Museum sign, "Cretaceous Forest," at the New Mexico Museum of Natural History and Science in Albuquerque stands in front of a grove of tropical trees in the museum atrium.*[13] *Museum sign reads, "The plants growing here, now found in tropic or wet areas around the world, are very similar to fossils found in New Mexico. Note the similarity between living plants and Cretaceous fossils, which are displayed along the balcony."*

ove: *Display at the Milwaukee Public Museum titled "The Lowland Forest Floodplain...A Home f*
n Dinosaurs" demonstrating plant life during the time of dinosaurs.[16] *To the inexperienced, these p*
eign, but in reality, all are living today. Some of these are non-flowering plants, such as the Dawn R
gingko, the ferns, and the bog (peat) moss. Others are flowering plants, such as the ground cover

B. Dawn Redwood
Metasequoia

A. Ginkgo Tree
Gingkophyta

C. Katsura Tree
Cercidiphyllum

D. Sycamore Tree
Platanus

E. Sensitive Fern
Onoclea

F. Bog Moss
Sphagnum

G. Pachysandra
Pachysandra

Above: *Living arrowhead plant* ***Sagittaria (species unidentified)****.*

Right: *Milwaukee Public Museum reconstruction of flowering arrowhead plant* ***Sagittaria (species unidentified)*** *based on fossils found at dinosaur site in Montana.*

Summary: Dinosaur-Era Plant Life

Examples of all of the major groups of plants living today have been found in dinosaur layers, including flowering plants and trees (angiosperms), plants without fruits or flowers (conifers, cycads, and ginkgos); vascular spore-forming plants (ferns, horsetails, and club mosses), and simple moss (peat moss).

My experiment predicted I would find modern plant and animal species in dinosaur rock layers if evolution was not valid. With the addition of plant life, my hypothesis has been confirmed.

Chapter 25

Coming Full Circle
My Conclusions

In the early days of paleontology, when 100-foot-long dinosaurs and huge flying reptiles were first discovered, paleontologists legitimately recognized that strange and uncommon animals existed in the past. These historical fossils soon became the showpiece for the theory of evolution — that animals changed dramatically over time, from one fossil rock layer to another and from past to present. Yet the basis for the major change of animals and plants over time hypothesis, which described what may have happened in the unobserved past, relied on the limited number of fossils collected up to that point in time. Now with advances in technology and expanded fossil collections, evolution scientists are discovering greater numbers of modern-appearing animals and plants in the dinosaur rock layers. This book brings this discussion to the next logical question: Did animals actually change over time, or did just some animals go extinct?

Limitations of This Book

I had only limited access to the fossils, was rarely allowed to touch them, and never had the opportunity to review the fossils under a microscope. Because of this limitation, I cannot be 100 percent certain that any one comparison presented in this book (of a dinosaur-era fossil and a modern living organism) is, in fact, the same genus or species. As time goes on, I may stand corrected on some of the comparisons I have made and future editions of this book will be updated accordingly.

This book is not exhaustive but is just a dim reflection of what is out there. With more time, money, access and assistance, I am confident I could add *thousands* of other examples of modern-appearing fossils since we photographed only 20,000 fossils, not the 200 million fossils that museums have collected.[1] I often wonder how much more complete this book would be if I was capable of photographing *all* of the fossils in the science museums of the world (perhaps 1,000 more institutions), *all* of the botanical garden leaf collections, and *all* of the modern organism collections in the natural history museums. I underscore my point with this example. We photographed a fossil that was identified as a hedgehog-like jaw from the dinosaur era, but I had access to only one of the known 16 species of hedgehogs living today for comparison. If I had greater access to a *range* of hedgehog skulls from each of these 16 species living today, including male and female and juvenile to adult, I believe an even closer match could have been made. The same could be said for all of the other animals and plants displayed in this book.

There were other limitations as well. There were discrepancies in the types of fossils museums displayed. Not once, at the 60 museums we visited, did I see a fossil of a modern type of bird found with a dinosaur. Yet, when I interviewed the scientists, they were aware of many examples. Similarly, even though nearly 100 complete mammal skeletons from the dinosaur era have been collected, I saw only two exhibited in museums.

Lastly, this book represents the efforts of just one couple carrying out this project in their spare time and at their own expense. Debbie and I continued working our normal jobs. No outside sources of financial support were available. Imagine if we had a $20 million grant from the National Science Foundation, a crew of 300 paleontologists and naturalists, and continued this project for another ten years.

Even though I have studied less that 0.01 percent of the worldwide fossil collections, it is significantly noteworthy that I have been able to identify modern-appearing dinosaur-era fossils from *all* seven major modern animal phyla living today, as well as fossils from *all* seven major modern plant divisions living today — fossils which appear similar to modern species. These examples are summarized on the next eight pages.

All Major Animal Phyla Living Today Also Found in Dinosaur Rock Layers

The Seven Major Animal Phyla Living Today[2]	Modern-Appearing Animals Found in Dinosaur Rock Layers*
Chordata (Vertebrates)	**Fish, Amphibians, Reptiles, Birds, Mammals** *-all three groups of fish living today* *-both major amphibian groups living today* *-all four reptile orders living today* *-"most or all" bird orders living today (by molecular divergence data)* *- all three types of mammals living today*
Echinodermata (Five-Sided Animals)	**Starfish, Brittle Stars, Sea Urchins, Sea Cucumbers, Crinoids** *-all five major classes of echinoderms living today*
Arthropoda (Animals with Exoskeleton)	**Insects, Shrimp, Crayfish, Lobsters, Crabs, Spiders, Scorpions, Centipedes, Millipedes** *-all of the major types of aquatic arthropods living today* *-all of the major insect orders living today* *-all four major arachnid orders living today* *-both major myriapod classes living today*
Mollusca (Shellfish)	**Oysters, Clams, Scallops, Mussels, Snails, Nautilus, Tusk Shells, Sea Cradles** *-all five major shellfish classes living today*
Annelida (Segmented Worms)	**Earthworms, Tube Worms** *-both major classes of annelids living today*
Porifera (Sponges)	**Glass Sponges, Demosponges, Bony Sponges** -all three classes of sponges living today
Cnidaria (Corals)	**Sea Pen, Humpback Coral** *-both hard and soft corals living today*

* See individual chapters for details of this chart

A Small Sample of Dinosaur-Era Animal Life

A Small Sample of Dinosaur-Era Aquatic Life
Styracosaurus

A Small Sample of Dinosaur-Era Plant Life

All Major Plant Divisions Living Today Also Found in Dinosaur Rock Layers

The Seven Major Plant Divisions Living Today[3]	Modern-Appearing Plants Found in Dinosaur Rock Layers*
Magnoliophyta (Flowering Plants)	Chestnut, Oak, Sassafras, Poplar, Walnut, Ash, Viburnum, Magnolia, Poppy, Rhododendron, Lily Pad, Sweetgum, Soapberry, Bay, Everglades Palm
Coniferae (Cone Trees)	Pine Tree, Sequoia, Redwood, Bald Cypress, Cook Pine
Pteridophyta (Spore-Forming Plants)	Tree Fern, Sensitive Fern, Wood Fern, Mosquito Fern, Shield Fern, Horsetail
Cycadophyta (Cycads)	Cycad, Coontie
Ginkgophyta	Ginkgo
Lycopodiophyta (Vascular Mosses)	Club Moss
Bryophyta (Avascular Mosses)	Peat Moss

* See individual chapters for details of this chart

My Conclusions

If I ignore the genus and species names and simply compare the fossils found in dinosaur rock layers to modern forms, I see a lack of significant change in all of the major animal phyla and all of the major plant divisions. My findings support the idea that animals and plants have not significantly changed (evolved) over time, but simply some animals and plants have gone extinct, while others have remained relatively unchanged.

The Bigger Picture

If living fossils were an isolated problem for the theory of evolution, they would be of less consequence. But, in the last two decades, other major obstacles for the modern theory of evolution have become more vocalized. Scientists on both sides of this cultural divide — a natural versus a supernatural origin of the universe and life itself — are familiar with these problems.

Problem 1: The Origin of the Universe

Astrophysicists who support the theory of the big bang have not been able to explain where the original matter and energy came from to form the universe.[4] Also, the laws of physics give nonsensical answers when they are applied to the big bang model.[4,5] One physicist, Dr. David Gross, recipient of the 2004 Nobel Prize in physics, described this situation as "total disaster."[4]

Problem 2: The Origin of Life

Scientists who support evolution have not been able to explain how the theoretical first form of life (a single-cell bacterium-like organism) could have formed spontaneously out of chemical elements. They have not been able to recreate the natural, spontaneous, formation of proteins, functional DNA, or functioning cell membranes — the three necessary components of all forms of life.

Problem 3: Traditional Adaptation Rejected by Evolution Scientists

Charles Darwin believed adaptation to be one of the mechanisms by which evolution occurred. His theory of adaptation implies that an animal was modified in direct response to the environment. Modern scientists do not believe that an individual multicellular animal can directly adapt to the environment and pass these changes to the next generation. They know this kind of adaptation is genetically impossible since there is no mechanism for an animal to sense a change in the environment using the body's skin cells, or nerve cells, and then pass this information to the DNA in the reproductive process. *The Encyclopedia of Evolution* recently suggested that the term, adaptation, should be removed from our scientific vocabulary.[6]

Problem 4: Theoretical Evolutionary Intermediates Absent for Most Organisms

There is a disproportionate number of claimed fossilized intermediates (or physical proofs of one animal changing into another) when compared to the vast number of fossils collected by museums. Darwin recognized that there were missing links when he wrote *The Origin of Species*, and devoted a two-chapter carefully worded apology concerning this situation.[7] He held out hope that as more fossils were found, these proofs of evolution (the intermediate animals that his theory predicted) would also be found. Now, 150 years after Darwin wrote his book, this problem still persists. Overall, the fossil record is rich — 200 million fossils in museums — but the predicted evolutionary ancestors are missing, seemingly contradicting evolution. For example: Museums have collected the fossil remains of 100,000 individual dinosaurs, but have not found a single direct ancestor for any dinosaur species. Approximately 200,000 fossil birds have been found, but ancestors for the oldest birds have yet to be discovered. The remains of 100,000 fossilized turtles have been collected by museums, yet the direct ancestors of turtles are missing.[8] Nearly 1,000 flying reptiles (pterosaurs) have been collected, but no ancestors showing ground reptiles evolving into flying reptiles have been found. Over 1,000 fossil bats have been collected by museums, but no ancestors have been found showing a ground mammal slowly evolving into a flying mammal. Approximately 500,000 fossil fish have been collected, and 100,000,000 invertebrates have been collected, but ancestors for the theoretical first fish, a series of fossils showing an invertebrate changing into a fish, are unknown. Over 1,000 fossil sea lions have been collected, but not a single ancestor of sea lions has been found. Nearly 5,000 fossilized seals have been collected, but not a single ancestor has been found. (See Volume I of this series for the interviews detailing these problems.)

Problem 5: The Three Best Fossil Examples of Evolution Problematic

The three best proofs of evolution using fossils (namely the evolution of birds from dinosaurs, the evolution of whales from a land mammal, and the evolution of man from apes) are problematic. Many of the evidences offered for these top three fossil proofs are erroneous. Here are some examples: (1) Scientists at the Milwaukee Public Museum in Wisconsin, the Jura Museum in Germany, and the Carnegie Museum in Pennsylvania placed scales on the head of the oldest bird, *Archaeopteryx*, to make it look like it had evolved from a reptile, yet no scales were found in the fossils[9]; (2) Scientists placed feathers on the dinosaurs *T. rex* and *Velociraptor* to make it appear that dinosaurs were evolving into birds, but no feathers were found on these fossils[10]; (3) The *National Geographic* "flying dinosaur," the best proof that birds evolved from dinosaurs, was a fake. Someone attached a dinosaur tail to a bird fossil to make it look like birds evolved from dinosaurs. When this discrepancy was pointed out to *National Geographic* before the story was released, they printed the story anyway[11]; (4) The world's foremost authority on whale evolution placed a whale's tail and flippers onto the drawings of a four-legged mammal, showing that whales evolved from a land mammal, even though he had not found the tail nor the flippers. He subsequently retracted this "best proof" of evolution while the video series for this book series was being filmed[12]. In all of these modern cases, the body parts in question were fabricated, sometimes innocently, sometimes maliciously. (See Volume I for the interviews detailing these most recent problems.) There are even more flagrant problems in the area of human evolution and these will be addressed in Volume III of this series.

Problem 6: Significance of Similarities Undermined

One of the foundational pillars for the theory of evolution is that similarities is evidence of evolution. Counting the number of similar features in different animals is one method by which evolution scientists construct evolution charts. For example, humans and apes have many similar features in their anatomy and this suggests to scientists who support evolution that humans evolved from apes.

This line of reasoning is being successfully challenged by scientists who oppose evolution. They point to a host of animals that are nearly identical in appearance yet belong to unrelated groups. For example, the appearance of a marsupial mouse (which, like a kangaroo, has an external pouch to raise and suckle its young) and a mouse (which does not have a pouch) are strikingly similar, yet one is a marsupial mammal and one is a placental mammal. Because of this, evolution scientists do not consider them closely related. The same is true for the marsupial mole and the placental mole. It is examples such as these that prompt scientists who oppose evolution to suggest that counting the number of similarities to construct evolution charts is precarious at best. (See Chapter 5 from Volume I of this series.)

The living fossils in this book also challenge this idea. Let me explain. If a scientist who supports evolution argues that two similar animals shown in this book are not closely related, such as the fossil and living nautilus, this undermines the very core of the evolution theory. If these two animals are unrelated, even though they have a plethora of similar features, then it undermines the evolutionary principle that animals with similar features are closely related.

Problem 7:
Best Evidences for Evolution
Eliminated over Time

Many of the historical "best evidences" for the theory of evolution have been eliminated over time. This is certainly true in my 30-year quest for an answer. All of the proofs for evolution that I was taught in my early adult years (the evidences that captured my allegiance to the theory of evolution), such as the big bang, Ontogeny Recapitulates Phylogeny, the spontaneous formation of a bacterium-like single-cell organism from chemicals (first life), the *National Geographic* outline of human evolution, and horse evolution, have all but been relegated to historical footnotes. This, in my opinion, is an unhealthy sign of a scientific theory. As each decade passes, the "best proofs" for evolution shift from one to another to another, while the earlier proofs are dismissed.

What Is *Your* Ending to This Story?

With these foundational problems, many would argue that the theory of evolution and the theory of the big bang need to be readdressed, altered, or abandoned. I certainly have grown to feel this way. These ideas have not reached a level of scientific clarity for me. But now it is your turn to weigh in with your opinion.

In the beginning of this book, I asked you to read my story with the caveat that *you* would furnish the ending. So what will it be? I'm curious to have *you* answer this question and complete this book: Has the theory of evolution been verified or has it been falsified by comparing life today to the fossil record? What do *you* think?

Appendix A: The Influence of Fossil Orientation

The manner in which fossils are displayed may aid in the perception that animals (or plants) changed dramatically over time or may aid in the perception that they did not change over time.

To demonstrate this, I will use a real example. Look at the two moon snails, one living and one from the dinosaur era, on Page 81. In your opinion, do they look like they could be the same species? Please commit to an answer before you read on.

Now look at the museum display below and compare the living snail (marked with an "X") to the dinosaur-era snail (marked with an "O"). Do they look like they are the same species? Again, please commit to an answer before proceeding.

The fact is both comparisons used the same fossil snail and the same living snail species, but the *orientation* of the fossil is different in the two examples. Let me explain.

When I originally viewed the museum display below (at the South Australian Museum), I noticed that the two snails "X" and "O" looked different, yet they were both labeled as moon snails. They were not turned in the same direction. Since I could not open the museum case and turn the specimens around to orient the specimens in the same direction, I had to wait and do it at home, electronically.

Using my computer, I rotated the photograph of the fossil marked "O" and placed it next to a photograph of a living moon snail (same species as the one in the cabinet), which I bought. Once the fossil and living snails were arranged the same way, they looked like the same species, as seen on Page 81. I marveled that simply rotating specimens they can be made to look more or less similar. In other words, the illusion of dramatic change can be created by the placement of the specimens in the museum cabinet.

Actual museum display demonstrating how these shellfish were placed in the case. Compare this display to page 81!

Appendix B: Use of Fossils Found below Dinosaur Layers

Most of the fossils in this book were found in one of the three recognized dinosaur rock layers: the Triassic, Jurassic, or the Cretaceous. However, I have included some fossils that are older (found below dinosaur layers), such as the sea cucumber in Chapter 4, the fossil club moss in Chapter 23, and the fossil sea limpets, daddy longlegs, sea pen and bony sponges in Chapter 25. My rationale for doing so follows.

Sometimes two similar fossils are found vertically separated from each other by hundreds of feet of rock, with no so-called similar fossil found in the rock layers between them. When this occurs, it appears the first organism was living long ago and then disappears, only to reappear later. Evolution scientists generally agree that these animals or plants did not go extinct but rather were not preserved in the layers between.

I offer an example with a living and a fossil coelacanth fish. This particular fish was last seen as a fossil in the dinosaur-era rock layers. Surprisingly, a living coelacanth fish was caught off the coast of South Africa in 1938! Evolution scientists generally believe the coelacanth fish had survived, remaining relatively unchanged, from the dinosaur-era until now. It simply was not preserved in the fossil record all these years.

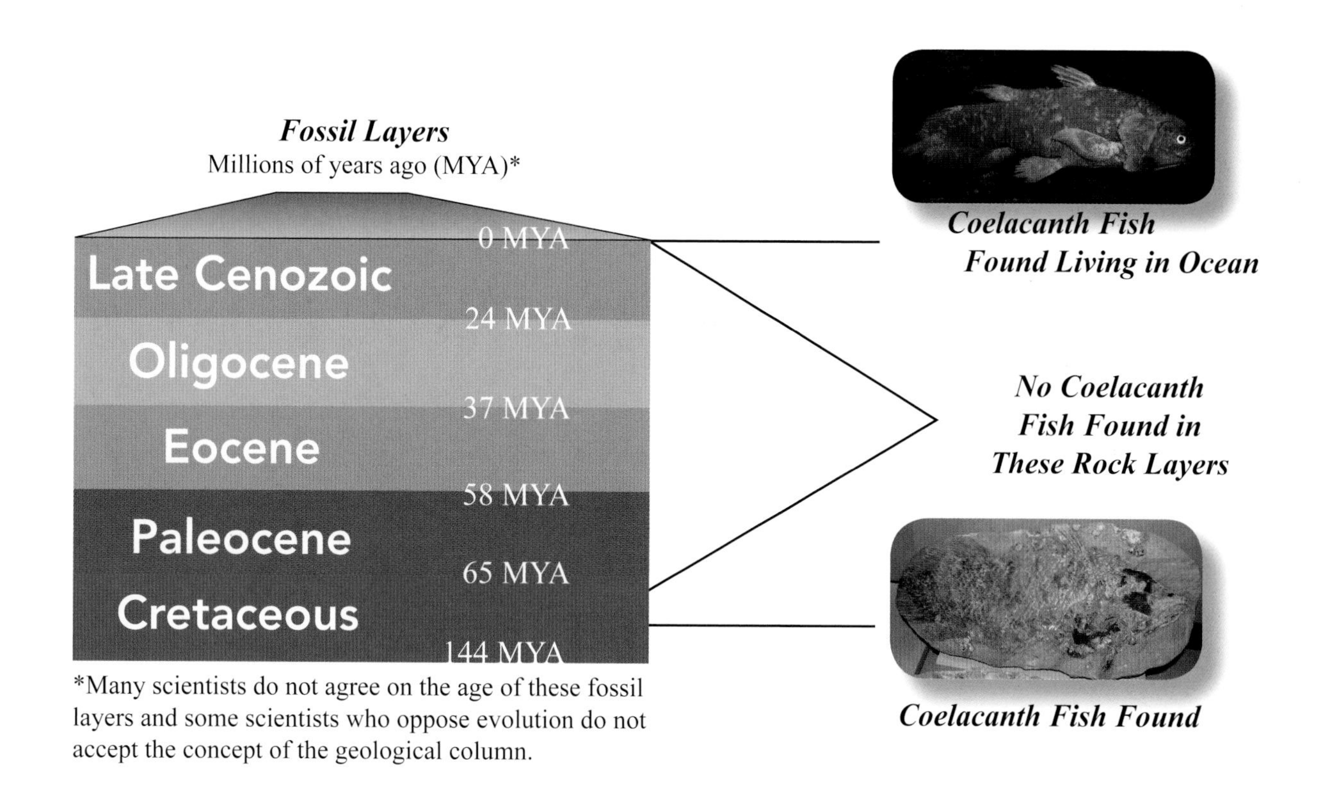

*Many scientists do not agree on the age of these fossil layers and some scientists who oppose evolution do not accept the concept of the geological column.

Applying These Principles to Other Fossils

I have applied this same principle to these other six fossils. Since the fossil sea pens, sea cucumbers, club moss, sea limpets, daddy longlegs and bony sponges were found below the dinosaur layers and since they are still living today, I have included them in this book based on the assumption that they "lived through" the years of the dinosaur era. In other words, they might not have been found (for the purposes of this book) in the dinosaur rock layers, but they were present in the dinosaur era and simply not preserved during that time.

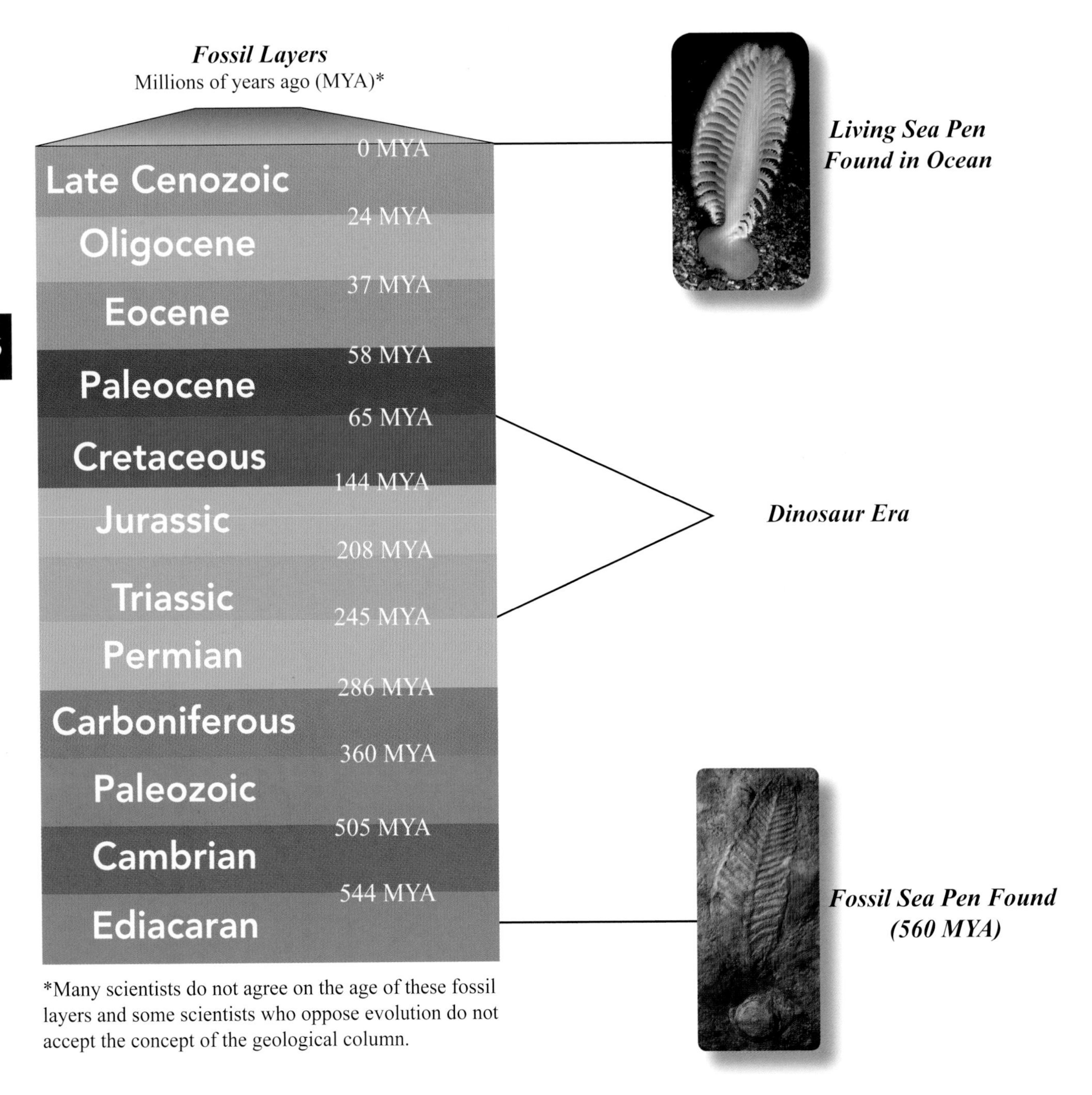

*Many scientists do not agree on the age of these fossil layers and some scientists who oppose evolution do not accept the concept of the geological column.

Appendix C: Difficulties in Vertebrate Identification

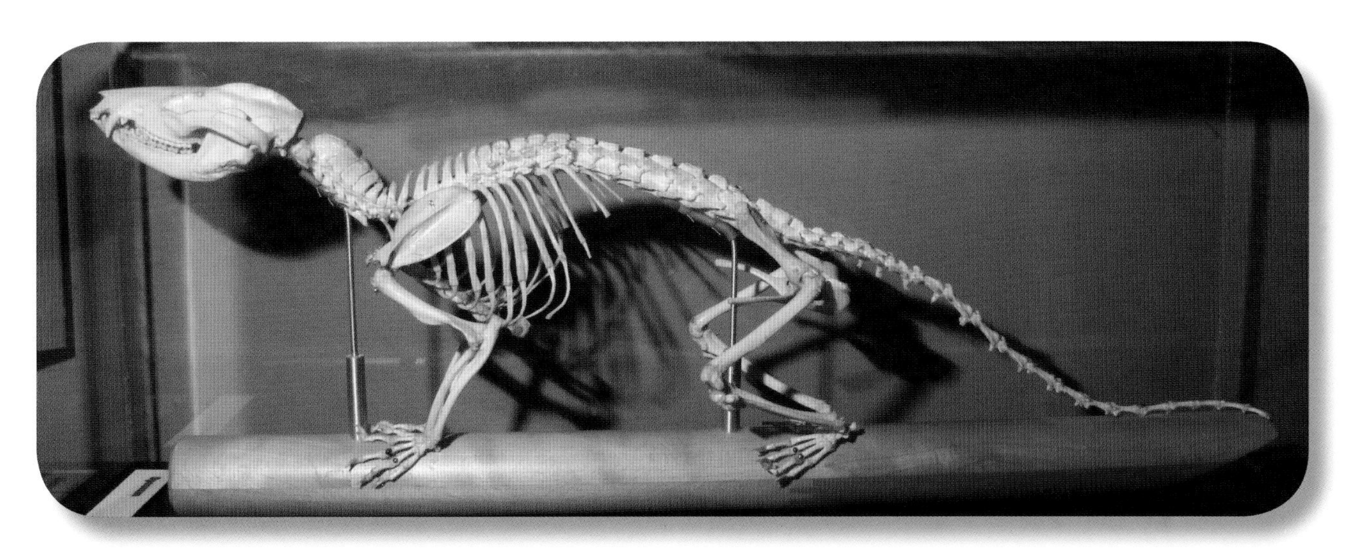

Can You Name These Five Animals?

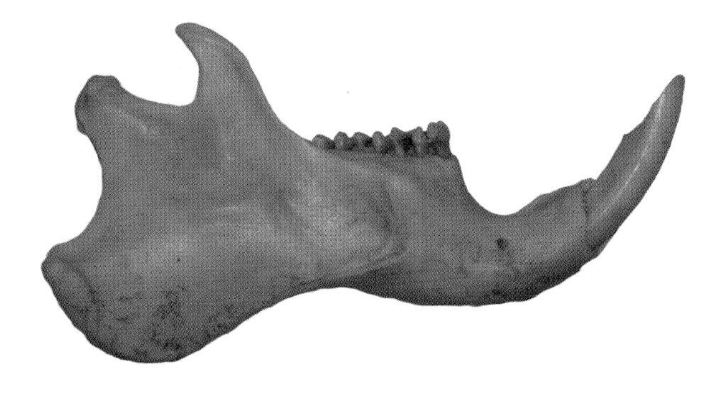

Most Are Unfamiliar with Vertebrate Anatomy

You most likely were able to identify the fossil starfish, crab, and lobster on the previous page, but I doubt you were able to identify the other two animals. Looking at these five pictures, I ask: Why are the three invertebrate animals (starfish, crab, and lobster) much easier to identify than the two vertebrate animals (those with a backbone)?

Invertebrates tend to fossilize with their natural covering preserved and therefore look nearly the same in the fossilized form as they do in real life.

On the other hand, when a vertebrate animal dies, the hair, skin, muscle, and cartilage tend to rot away before fossilization occurs. When this happens, there is usually nothing left but a pile of bones. The shape of the animal is not preserved nor the covering. Also, because the muscles and ligaments quickly rot away after death but before fossilization, the bones tend to get jumbled around. The leg bone is not necessarily connected to the foot bone, and the hand bone is not connected to the wrist bone, etc.

Because of this, we are generally dependent on a paleontologist, who frequently supports the theory of evolution, to tell us what the fossil vertebrate is. If you were not able to identify the reassembled Virginia possum's skeleton at the top of the previous page, how could you possibly identify a single tooth or bone of a modern mammal, such as the jawbone on the previous page?

This lack of ability for all but a few to recognize modern vertebrate bones or teeth could lead the public to conclude that modern-appearing vertebrates did not live during the time of the dinosaurs, especially when they are assigned such foreign names as *Steneosaurus* (Page 141) and *Gobiconodon* (Page 178).

Vertebrate Living Fossils Are Difficult to Identify in Museums

Appendix D: Assumptions in Fossil Reconstructions

Sometimes scientists make assumptions in reconstructions of fossils that are not readily apparent to the casual observer.

Color

Fossils do not preserve colors of the fur, feathers, or skin, nor do they preserve color patterns. One should be aware that when a color is selected by an artist for a reconstruction, it is simply an artistic choice and should be ignored. For example, the reconstruction of the dinosaur-era boa constrictor on Page 146 has a different color and pattern than the modern boa constrictor. Nothing can be concluded from this since these features are speculative. The same is true for the size and location of the spot near the tail of the bowfin reconstruction on Page 107. This should also be ignored.

The most obvious limitation in fossil reconstructions is the fur colors and patterns chosen for animal models. Compare the fur color and patterns on the *Gobiconodon* reconstructions B and C on Page 179. One has stripes, one does not. They do not look like the same animal, yet they are the same animal, interpreted by two different teams of artists/scientists.

Soft Tissues

If a scientist wants to make an animal appear more modern or less modern, he or she can (falsely) influence this by altering the non-bony anatomy, such as the cartilaginous nose and the cartilaginous ears, which are typically not preserved in fossils. Since these parts of an animal are not preserved, the scientist has great liberty in reconstructing them.

For example, look at the *Gobiconodon* reconstructions on Page 179 and notice how the artist — under a scientist's supervision — treated the ears. In reconstruction A, the ears are pointy and straight up. In model B, they are pointy, oriented backwards, yet lifted away from the body. In C, they are round, further down, and closer to the head. These three fossil reconstructions are based on the same fossils but are interpreted differently.

Body Positions

A scientist can also make any animal appear more or less modern by how he or she *positions* the bones. Turn to Pages 180-181 of *Volume I, Evolution: The Grand Experiment-The Quest for an Answer*. Here are side-by-side comparisons of fossil reconstructions of the same bird. In each example, the artist or scientist can make the fossil look more dinosaur-like by simply putting the wings out in front of the body or more bird-like by putting the wings out to the side. Body positioning, like fur color and soft tissues, must be ignored. You should only pay attention to the more substantive features — the bones — but as you will see, even these need to be viewed with a bit of skepticism.

Bones

When you see a fossil on display at a museum, you probably assume that the fossil bones are real. Regretfully, this is not always the case. Even in this book there are fossil bones that are not real. Look again at the *Gobiconodon* mammal skeleton on Page 178 in Chapter 21. Do you see anything wrong with this fossil skeleton? I certainly did not. How reliable does this fossil skeleton look to you? 100 percent reliable? 90 percent reliable? 20 percent reliable? When I saw this *Gobiconodon* skeleton at the Carnegie, I thought it was 100 percent reliable. After all, every bone was there, even the bones of the feet and tail. It wasn't until I got home and looked at the photos more closely that I realized there was a problem.

At the Carnegie Museum, beneath the *Gobiconodon* 'fossil' skeleton, there was a *drawing* of the same skeleton (see below).[2] Notice that in this drawing the bones are either filled in with dark pencil or just outlined with pencil. In this case, only the bones filled in with dark pencil are real.

Knowing this alters how you compare this dinosaur-era "skeleton" to the modern possum skeleton. Now you can ignore differences in the top of the skull, the neck bones, some of the backbones, and the scapula since these bones were missing.

The danger in all of this is that if a scientist or artist wants to make the composite skeleton look more archaic, he can avoid using modern-appearing bone shapes, and vice versa. As we pointed out in the "feathered dinosaur" reconstructions in Volume I, scientists should be obligated to reconstruct fossil vertebrates two ways: one interpretation showing how *modern* it could look and another interpretation showing how *ancient/extinct* it could look, depending on how the bones were substituted in various ways and positions. If this was carried out for each fossil vertebrate, the public would then view these interpretations much more skeptically.

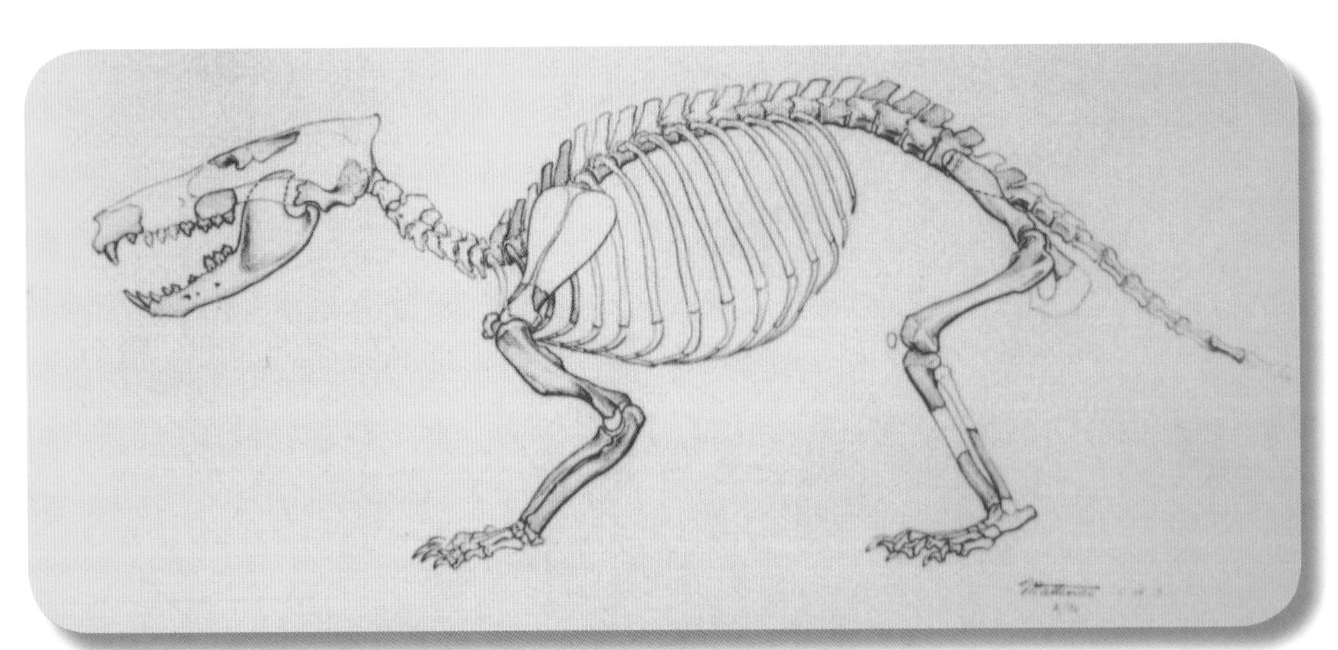

Above: *Drawing of the skeleton of Gobiconodon on display at the Carnegie Museum. The drawing shows which bones on display (Page 178) are real and which ones are not. Traditionally, scientists use diagrams with two different shades or colors to indicate which bones have been actually found.*

Appendix E: Further Analysis of Conclusions

Author's Note: *I will now present the anticipated arguments for and against evolution based on the information presented in this book.*

Modern Species Found/Not Found in Dinosaur Rock Layers

Point: Scientists who support the theory of evolution will point out that I have not found *any* modern species of plants or animals. They will suggest that although there are similarities between the fossils and the living organisms presented in this book, the differences between them are significant enough to justify naming these fossils with different genus and species names.

Counterpoint: Scientists who oppose the theory of evolution may suggest that the differences between the fossil and living plants and animals are not enough to justify changing genus or species names. They may even suggest that the naming of these fossils was motivated by the scientist's belief in the theory of evolution, not by actual significant differences between the fossil and the living organisms.

Not all Plants and Animals Found

Point: Scientists who support evolution will point out that I did not find *all* of the modern animal and plant types in the dinosaur layers, just some of them. To them, this proves that evolution occurred and that animals changed significantly.

Counterpoint: Scientists who oppose evolution will point out that I have presented evidence showing *all* of the major animal phyla and *all* of the major plant divisions living today were also alive at the time of the dinosaurs. They will point out that both theories, the theory of evolution and the theory of creation, have missing fossils. The theory of evolution, for example, has large gaps in the fossil record of bats, pterosaurs, dinosaurs, birds, whales, seals, sea lions, etc., and it is by "faith in the theory of evolution" that scientists fill in these gaps. The same is true for the theory of creation. For example, horses have not been found in dinosaur rock layers, nor have humans, or bats, or goats. The specific animals and plants living today that have *not* been found as fossils in the dinosaur layers are *implied* by the large number of other modern animals and plants that have been found.

Larger Mammals Not Found

Point: Scientists who support evolution will point out that most of the *larger mammals* living today, such as humans and horses and elephants, were not found in dinosaur rock layers only a few small and medium-sized mammals. They would suggest that this proves that evolution is true and that mammals changed dramatically over time.

Counterpoint: Scientists who oppose evolution may suggest that museums have been reluctant to display the complete mammal skeletons found in dinosaur layers. Without an openness to display these fossils, allowing all to photograph and publish them, it is impossible to evaluate this. It is imperative that the museums open up their collections to all, immediately.

Opponents of evolution may also suggest that if evolution is true then modern-appearing mammals, such as hedgehog-like, ringtail possum-like, beaver-like, Tasmanian Devil-like, duck-billed platypus-like, echidna-like, insectivore-like and shrew-like mammals should not be found in dinosaur rock layers. Finding modern-appearing mammals with dinosaurs is in direct opposition to the theory of evolution and is consistent with the idea of creation.

A

Angiosperms: Flowering plants. This group of plants includes roses, tomatoes, rhododendrons, the various grasses, and the flowering trees, such as sassafras, oak, palm, and apple.

***Archaeopteryx*:** This animal, according to some scientists, is one of the oldest known birds and had both bird-like and dinosaur-like features. Not all scientists agree.

B

Big Bang: The theory that the universe was created by a large explosion in space 10 to 20 billion years ago. The theory is based primarily on the observation of the redshift of starlight, which is interpreted to meant that the universe is expanding.

C

Copernicus, Nicolaus: 1473-1543. Polish astronomer who was one of several key scientists instrumental in overturning the idea that the earth is the center of our planetary system. He correctly deduced that the sun is.

Cretaceous: The most recent dinosaur fossil layer (also referred to as a geological time zone), the other two being the Triassic and Jurassic. Not all scientists agree on the age of these layers.

D

Darwin, Charles: 1809-1882. A British naturalist and evolution theorist. Charles Darwin did not invent the theory of evolution. His grandfather, Erasamus Darwin, and many others discussed and speculated about various evolutionary ideas. However, Charles Darwin was credited with providing a mechanism for evolution (natural selection), which he articulated in his book *The Origin of Species*. Not all scientists agree that his theory is valid.

DNA (deoxyribonucleic acid): A chemical compound in the cells which gives the commands regarding how to build the cell. DNA is made up of ribose sugar molecules with an oxygen atom removed, hence the name deoxyribose nucleic acid.

E

Echinoderm: An invertebrate animal with five-fold symmetry, meaning an animal without a backbone having five equal body parts. Echinoderms include starfish, brittle stars, sea cucumbers, sea lilies, and sea urchins.

Evolution: A scientific theory regarding how life came about. The theory of evolution suggests that all living plants and animals evolved from one another over long periods of time: Humans evolved from apes, apes evolved from monkeys, monkeys evolved from lemurs, lemurs evolved from non-primate mammals, mammals evolved from reptiles, reptiles evolved from amphibians, amphibians evolved from fish, fish evolved from invertebrates, invertebrates evolved from bacteria, and bacteria evolved from a single-cell organism which formed spontaneously in the ocean.

According to the theory, this process, from a single-cell organism to a human, took four billion years. Not all scientists agree that the theory is valid.

F

Fossil Record: The combined worldwide collection of known fossils. This collection tells the story of what kinds of animals and plants lived before the present time. Details such as where the fossils were collected and what layer they were found are frequently recorded. As more fossils are collected, the fossil record becomes more complete. Currently, over 200 million fossils have been collected by museums.

G

Galilei, Galileo: 1564-1642. Italian astronomer who championed the unpopular ideas of Nicolaus Copernicus. Both Galileo and Copernicus opposed the idea that the earth is the center of the planetary system. This false idea had been believed by scientists for nearly 2,000 years, since the time of Aristotle. [1]

I

Invertebrates: Any animal without a backbone, such as starfish, insects, and jellyfish.

J

Jurassic: The middle dinosaur fossil layer (also referred to as a geological time zone), the other two being the Triassic and Cretaceous. Not all scientists agree on the age of these layers.

K

Kingdom: The largest grouping of living organisms with similar traits. Examples include the animal kingdom, the plant kingdom, and the fungi kingdom.

L

Living Fossil: A fossil plant or animal which looks very similar to a living organism. [2]

M

Mesozoic Era: The time period of the dinosaurs which includes all three dinosaur fossil layers, including the Cretaceous, Jurassic, and Triassic.

Mutation: An accidental change in one or more letters in the DNA of a living organism. Most mutations are harmful to an organism. Some mutations have no effects at all. Rarely, a mutation can be be helpful, such as in sickle cell anemia *trait* and malaria. Sickle cell anemia frequently causes death.

N

Natural Selection: One of the suggested mechanisms for *how* evolution works. A large number of plants or animals would naturally display slight variations, such as smaller or larger size, lighter or darker color, etc. Darwin suggested that the most fit of these animals or plants would survive because of their ability to reproduce and survive a particular environment. The weaker would be killed off by nature, (i.e., natural selection). The killing off of weaker varieties is thought to have caused a species to continually improve or evolve.

A criticism of natural selection is that it fails to describe how animals could genetically change for the better, resulting in new or improved traits. Natural selection can only account for the removal of certain varieties but cannot, by itself, explain how new varieties or new traits came about. Darwin thought new varieties of plants or animals came about by the law of Acquired Characteristics, but this was proven wrong in 1889. New varieties of a species can only come about by accidental mutations. The question then becomes: Can accidental mutations cause a completely new type of animal to evolve, such as a dinosaur accidently mutating into a bird, over millions of years?

Nucleic Acid: DNA or RNA. See DNA and RNA.

O

Organic Chemicals: Chemical compounds which contain the element carbon. All living plants and animals contain organic chemicals.

Origin of Life: The theoretical event on earth when the very first form of life, a single-cell organism, came into being.

P

Pasteur, Louis: 1822-1895. French microbiologist and chemist who was instrumental in disproving spontaneous generation. The theory of spontaneous generation, which originated around the time of Aristotle (fourth century B.C.), was perpetuated for over 2,100 years. It was considered an "article of faith" by many biologists until it was finally disproved by Pasteur in 1859.

Phylum: The largest grouping of animals with similar traits. Examples include the worms, corals, sponges, arthropods, molluscs, and echinoderms. From largest to smallest, all living organisms can be grouped by kingdom, phylum, class, order, family, genus, and species.

Protein: A chain of amino acids. Proteins have many functions in a cell, such as catalyzing (speeding up) chemical reactions, copying DNA, and forming essential structures in the cell.

Proteinoid: An unnatural organic compound brought about by heating dried, purified amino acids. A proteinoid does not have the normal bonds between amino acids that a protein has, and has limited functions. Proteinoids are theorized by some evolution scientists to be the precursors to proteins, but proteinoids have never been observed to form in nature and have never been observed to convert to proteins.

R

Redi, Francesco: 1626-1697. Italian physician and scientist who was instrumental in disproving one of the tenets of the theory of spontaneous generation — maggots arising from rotting meat. He showed that maggots growing on rotting meat did not represent the spontaneous generation of life, but rather the contamination of the meat by flies. Maggots are the larval stage of flies.

RNA (ribonucleic acid): A nucleic acid. RNA is made up of ribose sugar molecules, hence the name ribo(se)nucleic acid (RNA).

S

Species: A group of organisms capable of interbreeding and producing *fertile* offspring.

Spontaneous Generation: The disproved theory that animals came about spontaneously, over short periods of time. This theory suggested that maggots developed out of decaying meat over a period of two weeks and that mice developed from dirty underwear over a period of three weeks. The last vestiges of this theory were disproved by the work of Louis Pasteur in 1859.

T

Triassic: The oldest, or first, dinosaur fossil layer (also referred to as a geological time zone), the other two being the Jurassic and Cretaceous. Not all scientists agree on the age of these layers.

V

Vertebrates: Animals with backbones, such as human beings, frogs, and fish.

Chapter 1: The Challenge That Would Change My Life

1. Milner, R. (1990). The Encyclopedia of Evolution: Humanity's Search for Its Origins. New York: Facts on File Publishers. p. 206.

2. Richardson, M, (1997). *Anatomy and Embryology*, There is no highly conserved embryonic stage in the vertebrates: implications for current theories of evolution and development, August 1997, pp. 91-106.

3. Grigg, R. (1996). *Creation*, Ernst Haeckel Evangelist for evolution and apostle of deceit, March 1996, pp. 33-36.

4. Grigg, R. (1998). *Creation*, Fraud rediscovered, March 1998, pp49-51.

5. Interview with Dr. Daniel Gasman, professor of history at John Jay College of Criminal Justice and professor of history at the Graduate Center at the City University of New York, for video series *Evolution: The Grand Experiment*, conducted May 21, 2004, by author. Dr. Gasman is the author of *The Scientific Origins of National Socialism* (New York: American Elsevier, 1971) and also *Haeckel's Monism and the Birth of Fascist Ideology* (publisher P. Lang, 1998). Dr. Gasman is considered an expert on Haeckel.

6. Cort, J. and J. McMaster, (producers and directors). (2003). *The Elegant Universe: Welcome to the 11th Dimension* (Episode 3), shown on PBS. [Nova DVD Series.] (Available from WGBH Boston Video call 888 255-9231, fax 802 864-9846 and online at www.shop.wgbh.org.)

BRIAN GREENE: (Professor of Physics and Mathematics at Columbia University and author of *The Elegant Universe*): *"There's always been a couple of problems with the big bang theory. First, when you squeeze the entire universe into an infinitesimally small, but stupendously dense package, at a certain point, our laws of physics simply break down. They just don't make sense anymore."*

DAVID GROSS (University of California, Santa Barbara): *"The formulas we use start giving answers that are nonsensical. We find total disaster. Everything breaks down, and we're stuck."*

BRIAN GREENE: *"And on top of this, there's the bang itself. What exactly is that?"*

ALAN GUTH (Massachusetts Institute of Technology): *"That's actually a problem. The classic form of the big bang theory really says nothing about what banged, what happened before it banged, or what caused it to bang."*

BRIAN GREENE: *"Refinements to the big bang theory do suggest explanations for the bang, but none of them turn the clock back completely to the moment when everything started."*

PAUL STEINHARDT (Princeton University): *"Most people come at this with the naïve notion that there was a beginning — that somehow space and time emerged from nothingness into somethingness."*

BURT OVRUT (University of Pennsylvania): *"Well, I don't know about you, but I don't like nothing. Do I really believe that the universe was a big bang out of nothing? And I'm not a philosopher, so I won't say. But I could imagine to a philosopher, that is a problem. But to a physicist, I think, it's also a problem."*

BRIAN GREENE: *"Everyone admits there are problems."*

7. Interview with Dr. Duane Gish, biochemist, Institute for Creation Research, Santee, California, conducted in February 1998 by author for video series *Evolution: The Grand Experiment.*

Chapter 2: How Can You Verify Evolution?

1. See Chapter 1, footnote #2.

2. Lerner, Eric. (2004). *New Scientist,* Bucking the Big Bang, May 22, 2004, p. 2448.

3. *Scientific Method* (n.d.). Retrieved July 20, 2008, from http://en.wikipedia.org/wiki/Scientific_method. "Scientific method refers to the body of techniques for investigating phenomena, acquiring new knowledge, or correcting and integrating previous knowledge. It is based on gathering observable, empirical and measurable evidence subject to specific principles of reasoning. A scientific method consists of the collection of data through observation and experimentation, and the formulation and testing of hypotheses."

4. Strickberger, M. (1996). *Evolution,* (2nd Edition). Boston: Jones and Bartlett Publishers. p. 604. "Living fossil: An existing species whose similarity to ancient ancestral species indicates that very few morphological changes have occurred over a long period of geological time."

Chapter 3: The Naming Game

1. *Species* (n.d.). Retrieved July 20, 2008, from http://en.wikipedia.org/wiki/Species. "A species is often defined as a group of organisms capable of interbreeding and producing fertile offspring. While in many

cases this definition is adequate, more precise or differing measures are often used, such as based on similarity of DNA or morphology. Presence of specific locally adapted traits may further subdivide species into subspecies."

2. Boule, M. (1923). *Fossil Men: Elements of Human Paleontology,* (2nd French Edition). London: Oliver and Boyd Publishers. p. 72.

3. Boule, M. and H. Vallois. (1957). *Fossil Men.* New York: The Dryden Press. pp. 344-345.

4. Live oyster species, *Crassostrea virginica*, identified by Paul Cook and Mr. Patrick Banks, oyster program manager for Louisiana Department of Wildlife and Fisheries.

5. Interview with Dr. Thomas Williamson, curator of paleontology, New Mexico Museum of Natural History and Science, Albuquerque, conducted March 19, 2002, by author for video series *Evolution: The Grand Experiment.*

6. Dinosaur-era snail in Chapter 8 initially identified as *Triasaminocola (species undetermined).* Later name changed to *Lioplacodes (species undetermined) and* identified by Bill Parker, vertebrate paleontologist, Division of Resource Management, Petrified Forest National Park..

7. Dinosaur-era bowfin in Chapter 12 initially identified by museum staff as *Amia (species unidentified).* Later name changed to *Kindleia fragosa* by Dr. Peter Sheehan, head of geology department at the Milwaukee Public Museum, adjunct professor, Department of Geosciences at the University of Wisconsin, Milwaukee.

8. Dinosaur-era Sensitive Fern in Chapter 23 initially identified by Milwaukee Public Museum staff as *Onoclea (species unidentified).* Later name changed to *Hydropteris pinnata* by Dr. Peter Sheehan, head of geology department at the Milwaukee Public Museum, adjunct professor, Department of Geosciences at the University of Wisconsin, Milwaukee.

9. Dinosaur-era fern in Chapter 23 initially identified as *Laccopteris smithii.* Later name changed to *Phlebopteris smithii.*

10. Dinosaur-era lily in Chapter 24 initially identified by Milwaukee Public Museum staff as *Nympheta.* Later name changed to *Nelumbium montanum* by Dr. Peter Sheehan, head of geology department at the Milwaukee Public Museum, adjunct professor, Department of Geosciences at the University of Wisconsin, Milwaukee.

11. Joyce, Walter G. (July 2000). The First Complete Skeleton of Solnhofia parsoni (Crytodira, Eurysternidae) from the upper Jurassic of Germany and its taxonomic implications. *Journal of Paleontology, Volume 74, Issue 4,* pp. 684–700. "Synonymies of Solnhofia with such turtle genera as Eurysternum, Idiochelys, Plesiochelys, Thalassemys, and Euryaspis can now be refuted. Similarities with Hydropelta are apparent, but not considered sufficient to support a synonymy. "

12. Living magnolia Chapter 24 initally called *Manglietia yunnanensis.* Later name changed to *Magnolia yunnanensis* and identified by Dennis Werner, JC Raulston Arboretum, North Carolina State University.

Chapter 4: Echinoderms

1. Live brittle star, *Ophiopholis,* genus identification by Dr. Gordon Hendler, curator of echinoderms, Natural History Museum of Los Angeles County, and by Dr. Robert Stone, National Oceanic and Atmospheric Administration (NOAA) Fisheries, Auke Bay Laboratories, Juneau, Alaska.

2. *Echinoderm* (n.d.). Retrieved March 25, 2008, from http://en.wikipedia.org/wiki/Echinoderm. "Two main subdivisions of Echinoderms are traditionally recognised: the more familiar, motile Eleutherozoa, which encompasses the *Asteroidea* (starfish), *Ophiuroidea* (brittle stars), *Echinoidea* (sea urchin and sand dollar) and *Holothuroidea* (sea cucumbers); and the sessile *Pelmatazoa*, which consist of the crinoids. Some crinoids, the feather stars, have secondarily re-evolved a free-living lifestyle. A fifth class of Eleutherozoa, consisting of just two species, the Concentricycloidea (sea daisies), were recently [1] merged into the Asteroidea. The fossil record contains a host of other classes which do not appear to fall into any extant crown group."

Chapter 5: Aquatic Arthropods

1. Live shrimp, *Litopenaeus setiferus*, species identification by Paul Cook and Martin Bourgeois, shrimp program manager for Louisiana Department of Wildlife and Fisheries.

2. Live prawn, *Macrobrachrium rosenbergii*, species identification by Leonard Sonnenschein, president of the World Aquarium and Conservation for the Oceans Foundation, St. Louis, Missouri, USA.

3. Live crayfish, *Procambrus clarkii*, species identification by Paul Cook and Jody David, crawfish

project manager for Louisiana Department of Wildlife and Fisheries.

4. *Arthropod* (n.d.). Retrieved July 25, 2008, from http://en.wikipedia.org/wiki/Arthropod.

"**Trilobites** are a group of formerly numerous marine animals that disappeared in the Permian-Triassic extinction event, though they were in decline prior to this killing blow, having been reduced to one order in the Late Devonian extinction.

Chelicerates include spiders, mites, scorpions and related organisms. They are characterised by the presence of chelicerae.

Myriapods comprise millipedes and centipedes and their relatives and have many body segments, each bearing one or two pairs of legs. They are sometimes grouped with the hexapods.

Hexapods comprise insects and three small orders of insect-like animals with six thoracic legs. They are sometimes grouped with the myriapods, in a group called Uniramia, though genetic evidence tends to support a closer relationship between hexapods and crustaceans.

Crustaceans are primarily aquatic (a notable exception being woodlice) and are characterised by having biramous appendages. They include lobsters, crabs, barnacles, crayfish, shrimp and many others."

Author's Note: Arthropods are typically classified into the above mentioned five groups: (1) Trilobites (*extinct* aquatic artropods); (2) Chelicerates (spiders and mites, which are non-aquatic); (3) Myriapods, millipedes, and centipedes, which are non-aquatic; (4) Hexapods (non-aquatic insects); and (5) Crustaceans. Other than crustaceans, the only other notable aquatic arthropod is the horseshoe crab, which is a member of the Chelicerates.

5. *Crustacea*, (n.d.). Retrieved March 25, 2008, from http://en.wikipedia.org/wiki/Crustacean. Six classes of crustaceans are generally recognized. They are:

Branchiopoda — including brine shrimp (Artemia) and Triops (Notostraca)

Remipedia — a small class restricted to deep caves connected to salt water, called anchialine caves

Cephalocarida — horseshoe shrimp

Maxillopoda — various groups, including barnacles and copepods. It contains Mystacocarida and Branchiura, which are sometimes treated as their own classes

Ostracoda — small animals with bivalve shells

Malacostraca — the largest class, with the largest and most familiar animals, such as crabs, lobsters, shrimp, krill, and [terrestrial] woodlice.

Author's Note: I have arbitrarily classified the familiar groups of crustaceans, such as shrimp, crayfish, fresh water prawns, lobsters, and crabs, plus horseshoe crabs, as major groups of aquatic arthropods and have arbitrarily classified these less familiar and diminutive varieties, such as Branchiopoda, Remipedia, Cephalocarida, Maxillopoda, Ostracoda, and krill, as "minor." (Most krill are 1 to 2 cm long as adults.)

Chapter 6: Land Arthropods

1. Live dragonfly, *Pachydiplax longipennis*, species identification by Paul Cook and Michael Seymour, non-game biologist for Louisiana Department of Wildlife and Fisheries.

2. Live katydid, *Scudderia furcata*, species identification by Edward Spevak, curator of invertebrates, St. Louis Zoo, Missouri.

3. Live water skater, *Gerris remigis*, species identification by Edward Spevak, curator of invertebrates, St. Louis Zoo, Missouri.

4. Fossil mayfly larvae, *Australurus plexus*, identified by museum paleontology staff and Kirsty Marshall, information officer, Discovery Center, Museum Victoria, Melbourne, Australia.

5. Live cricket, *Acheta domesticus,* species identification by Edward Spevak, curator of invertebrates, St. Louis Zoo, Missouri.

6. **Author's Note:** Listed are the *major* insect orders living today that have been found in dinosaur rock layers or below dinosaur rock layers:

-Butterflies (Class Insecta, Order Lepidoptera)

-Mayflies (Class Insecta, Order Ephemeroptera)

-Dragonflies (Class Insecta, Order Odonata)

-Water Skaters (Class Insecta, Order Hemiptera)

-Crickets (Class Insecta, Order Orthoptera)

-Termites (Class Insecta, Order Isoptera)

-Roaches (Class Insecta, Order Dictyoptera)

-Scorpionflies (Class Insecta, Order Mecoptera)

-Beetles (Class Insecta, Order Coleoptera)

-Bees (Class Insecta, Order Hymenoptera)

-Flies (Class Insecta, Order Diptera). Retrieved August 23, 2005, from www.ncsu.edu/unity/lockers/ftp/bwiegman/fly_html/diptera.html#TOC1. "The earliest fossil flies are known from the Upper Triassic of the

Mesozoic geological period, some 225 million years ago (Evenhuis, 1995)."

Author's Note: According to the principles outlined in Appendix B, if an animal is found in fossil layers below dinosaurs and if it lives today, then presumably it also lived during the time of the dinosaurs.

-Stone Flies (Class Insecta, Order Plecoptera). See *The Stone Flies (Plecoptera)* (n.d.). Retrieved August 23, 2005, from www.earthlife.net/insects/plecopt.html.

Author's Note: Article states the earliest known stone flies found as fossils are from Permian. According to the principles outlined in Appendix B, if an animal is found in fossil layers below dinosaurs and if it lives today, then presumably it also lived during the time of the dinosaurs.

-Lacewings (Class Insecta, Order Neuroptera). See *Insect Fossils* (n.d.). Retrieved August 23, 2005 from Kendall Bioresearch Services at www.kendall-bioresearch.co.uk/fossil.htm.

Author's Note: Site indicates that lace wings have been found in Permian. According to the principles outlined in Appendix B, if an animal is found in fossil layers below dinosaurs and if it lives today, then presumably it also lived during the time of the dinosaurs.

I have arbitrarily considered these diminutive or relatively unknown insect orders as *minor*: (1) Thyansura (three-pronged bristletails); (2) Diplura (two-pronged bristletails); (3) Protura (Proturans); (4) Collembola (springtails); (5) Dermaptera (earwigs); (6) Psocoptera (book and bark lice); (7) Mallophaga (biting lice); (8) Siphunculata (sucking lice); (9) Thysanoptera (thirps); (10) Siphonoptera (fleas); (11) Strepsiptera (stylops); (12)Trichoptera (Caddis flies); (13) Grylloblatodea (ice bugs); (14) Embioptera (web spinners); (15) Zoraptera (less than 3mm long); and (16) Stylops (4mm long).

7. Interview with Dr. William Clemens, professor of integrated biology and curator of the Museum of Paleontology, University of California, Berkeley, conducted November 9, 1998, by author for video series *Evolution: The Grand Experiment*.

8. Long, J. (1995). *The Rise of Fishes: 500 Million Years of Evolution.* Baltimore: Johns Hopkins University Press. p. 146.

9. **Author's Note:** There are four classes of the subphylum Myriapoda: Chilopoda (centipedes), Diploda (millipedes), Pauropoda (pauropods, pauropodans, and progoneates), and Symphyla (pseudocentipedes and symphylans). I have arbitrarily assigned the last two classes to the minor group since few have heard of or have personal knowledge of these more obscure members of Myriapoda.

10. *Arthropod* (n.d.). Retrieved August 23, 2005, from http://en.wikipedia.org/wiki/Arthropod. "Myriapods comprise millipedes and centipedes and their relatives and have many body segments, each bearing one or two pairs of legs. They are sometimes grouped with the hexapods."

11. **Author's Note:** There are 11 living orders of arachnids, and I have classified them as major or minor based on their size and/or obscurity.

Major **Orders of Arachnida**

Acarina: Mites and Ticks

Araneae: Spiders

Scorpiones: Scorpion

Opiliones: Daddy longlegs. See *Arachnid* (n.d.). Retrieved March 26, 2008, from http://en.wikipedia.org/wiki/Arachnid#cite_note-6. "Opiliones (better known as "harvestmen" or "daddy longlegs")...have been found in the 400-million-year-old Rhynie cherts of Scotland, which look surprisingly modern, indicating that the basic structure of the harvestmen hasn't changed much since then."

Author's Note: According to the principles outlined in Appendix B, if an animal is found in fossil layers below dinosaurs and if it lives today, then presumably it also lived during the time of the dinosaurs.

Minor **Obscure Orders of Arachnida**

Amblypygi: Tailless whip scorpions and cave spiders

Palpigradi: Microwhip scorpions

Pseudoscorpionida: Pseudoscorpions (8mm long)

Ricinulei: Hooded tickspiders

Schizomida: Eyeless arthropods in soil

Solifugae: Camel spiders, wind scorpions, sun spiders

Uropygi: Whip scorpions

Chapter 7: Bivalve Shellfish

1. Fossil freshwater clam, *Elliptio (species undetermined),* genus identified by Bill Parker, vertebrate paleontologist, Division of Resource Management, Petrified Forest National Park.

Chapter 8: Snails

1. Snail, *Lioplacodes (species undetermined),* genus identified by Bill Parker, vertebrate paleontologist, Division of Resource Management, Petrified Forest National Park.

Chapter 9: Other Types of Shellfish

1. Geyer, G. and Kukla, P. (March 1990). "An Unusual New Lepidopleurid Polyplacophoran (Mollusca) from the Germanic Muschelkalk (Triassic)." *Journal of Paleontology, Vol. 64, No. 2*, pp. 222-228.

2. *Mollusca* (n.d.). Retrieved March 28, 2008, from http://en.wikipedia.org/wiki/Mollusc.

Author's Note: Article states that there are seven living classes of mollusca: (1) Class Bivalves (clams, oysters, scallops, mussels); (2) Class Gastropoda (snails); (3) Class Scaphopoda (tusk shells); (4) Class Cephalopoda (nautilus); (5) Class Polyplacophora (chitons); (6) Class Monoplacophora (deep-sea limpet-like creatures), (7) Class Aplacophora (deep-sea wormlike creatures and sub-class Caudofoveata, deep-sea wormlike creatures now generally recognized as a subclass of Aplacophora).

Author's Note: I have arbitrarily relegated the last two obscure classes as "minor."

3. *Monoplacophora* (n.d.). Retrieved March 26, 2008, from http://en.wikipedia.org/wiki/Monoplacophora. "Monoplacophora is a class or, more likely, polyphyletic group of shelled mollusks (the name 'Monoplacophora' means 'bearing one plate'). These organisms were known only from the fossil record, and thought to have vanished in Devonian times, until in April 1952 a living specimen was dredged up from deep marine sediments in the Middle America Trench off the Pacific coast of Costa Rica. It was named Neopilina galatheae by its discoverer, Danish biologist Dr. Henning Lemche. So far, more than two dozen living species of Monoplacophora have been discovered; the first to be photographed live was *Vema hyalina*, at a depth of 400 m off Catalina Island, California, in 1977. All the present species live deep down in ocean trenches. An attempt at a common name, gastroverm, has proved unsuccessful."

Author's Note: Article indicated that Monoplacophora have been found in Devonian fossil layers. According to the principles outlined in Appendix B, if an animal is found in fossil layers below dinosaurs and if it lives today, then presumably it also lived during the time of the dinosaurs.

Chapter 10: Worms

1. Rouse, G., Pleijel, F. and McHugh, D. (2002). *Annelida.* Retrieved July 20, 2008, from http://www.tolweb.org/Annelida. "Through most of the 20th century Annelida was split into three major groups: Polychaeta, Oligochaeta (earthworms etc.), and Hirudinea (leeches)....It is now recognized that Oligochaeta and Hirudinea, comprised of several thousand species, form a clade and should be referred to either as Oligochaeta (Siddall et al., 2001) or Clitellata (Martin, 2001)." **Author's note:** Christmas tree worms are members of the class Polychaeta. Earthworms are members of the class Oligochaeta.

Chapter 11: Sponges and Corals

1. *Sponges* (n.d.). Retrieved March 28, 2008, from Wikipedia, http://en.wikipedia.org/wiki/Sponge#Taxonomy. "Sponges are divided into classes based on the type of spicules in their skeleton. The three classes of sponges are bony (Calcarea), glass (Hexactenellida), and spongin (Demospongiae). Some taxonomists have suggested a fourth class, Sclerospongiae, of coralline sponges, but the modern consensus is that coralline sponges have arisen several times and are not closely related."

2. *Calcarea* (n.d.). Retrieved August 29, 2005, from http://www.ucmp.berkeley.edu/porifera/calcarea.html. "The Calcarea first appears at the base of the Lower Cambrian and has persisted until the present. Greater than 100 fossil genera are known."

Author's Note: Article indicates that Calcarea sponges have been found in Cambrian. According to the principles outlined in Appendix B, if an animal is found in fossil layers below dinosaurs and if it lives today, then presumably it also lived during the time of the dinosaurs.

3. **Author's Note:** Examples of *both* subclasses of corals living today have been found, including soft corals such as sea pens (Class Anthozoa, Subclass Alcyonaria-Octocorallia), and hard corals such as humpback coral (Class Anthozoa, Subclass Zoantharia Hexacorallia). For sea pens, see discussion in Appendix B: Use of Fossils Found below Dinosaur Layers.

Chapter 12: Bony Fish

1. Sturgeon, *Acipenser ablertensis,* species identified by Fred Hammer, education coordinator at Field Station Visitor Center, Dinosaur Provincial Park, Alberta, Canada.

2. Long, J. (1995). *The Rise of Fishes: 500 Million Years of Evolution.* Baltimore: Johns Hopkins University Press. p. 180.

3. Interview with Dr. David Weishampel, anatomist and paleontologist, Johns Hopkins University and lead editor of the encyclopedic reference book *The Dinosauria*, conducted November 16, 1998, by author for video series *Evolution: The Grand Experiment.*

4. Interview with Dr. John Long, paleontologist and head of science, Museum Victoria, Melbourne, Australia, for video series *Evolution: The Grand Experiment*, conducted March 2005.

5. Long, J. (1995). *The Rise of Fishes: 500 Million Years of Evolution.* Baltimore: Johns Hopkins University Press. p. 162.

6. Dinosaur-era bowfin, *Kindleia fragosa,* from Hell Creek Formation, on display at the Milwaukee Public Museum, identified by Dr. Peter Sheehan, head of Geology Department at the Milwaukee Public Museum, adjunct professor, Department of Geosciences at the University of Wisconsin, Milwaukee.

7. Interview with Dr. William Clemens, professor of integrated biology, curator of the Museum of Paleontology, University of California, Berkeley, conducted November 9, 1998, by author for video series *Evolution: The Grand Experiment.*

8. Quote from display at The New Mexico Museum of Natural History and Science in Albuquerque (March 19, 2002). "Marine life in New Mexico's Cretaceous seas was similar to that in present-day oceans. Modern groups of sharks and rays appeared and flourished in the Cretaceous. Some fossil forms are almost identical to living species. Many bony fishes, such as flounder, lived 75 million years ago."

Chapter 14: Jawless Fish

1. Long, J., (1995). *The Rise of Fishes: 500 Million Years of Evolution.* Baltimore: Johns Hopkins University Press. p. 44.

Chapter 15: Amphibians

1. *Karaurus* (n.d.). Retrieved July 20, 2008, from http://en.wikipedia.org/wiki/Karaurus.

2. Interview with Dr. Brint Breithaupt, director, University of Wyoming Geological Museum, conducted July 25, 1997, by author for video series *Evolution: The Grand Experiment.*

3. **Author's note:** There are three types of amphibians living today: frogs, salamanders, and legless salamanders. I have arbitrarily classified the more familiar groups, frogs and salamanders, as major groups and the obscure group of legless salamanders as minor. Even so, legless salamanders such as Eocaecilia have also been found in dinosaur rock layers. See www.palaeos.com/Vertebrates/Units/Unit180/180.100.html, accessed September 8, 2008.

Chapter 16: Crocodilians

1. Dinosaur-era alligator, *Albertochampsa langstoni,* species identified by Fred Hammer, education coordinator at Field Station Visitor Center, Dinosaur Provincial Park, Alberta, Canada.

2. Dinosaur-era crocodile, *Leidyosuchus canadensis,* species identified by Fred Hammer, education coordinator at Field Station Visitor Center, Dinosaur Provincial Park, Alberta, Canada.

Chapter 17: Snakes

1. Rage, J., Guntupalli, Prassad, G.V.R, and Bajpai, S., (2004). *Cretaceous Research*, Volume 25, Issue 3, Pages 425-434. *Additional snakes from the uppermost Cretaceous (Maastrichtian) of India.* Retrieved March 5, 2008, from www.sciencedirect.com. "Thus far, snakes from the Mesozoic of India are known only from the Maastrichtian."

2. Fraser, N.C. and Sues, H.D. (1994). *In the Shadow of the Dinosaurs: Early Mesozoic Tetrapods.* Cambridge University Press, pp. 30-31. "The earliest snake fossils, three vertebrae identified as *Lapparentophis defrennei* Hoffstetter, come from the Lower Cretaceous of North Africa (Rage, 1984)."

3. *Madtsoiidae* (n.d.). Retrieved March 5, 2008, from http://en.wikipedia.org/wiki/Madtsoiidae.

4. Baucho, R. (2006). *Snakes: A Natural History.* New York: Sterling Publishing Company, Inc. p. 29. "Evolution of Snakes. Paleontology tells us little about snakes, first evolutionary steps. The oldest fossils, found in Algeria, in the middle Cretaceous deposits (100 to 96 million years old) consist of a few damaged vertebrae. Another Algerian fossil, *Lapparentophus defrennes,* of which we have only very partial remains, is either a contemporary of the former or slightly more recent. *Paryachis prolematicus*, on the other hand, a complete fossil about 3 1/2 ft. (1M) long, from the middle Cretaceous, about 94 to 97 million years old, was found in the Middle East. Although it was serpentiform, it probably had very minimal hind limbs. Despite its remarkable state of

preservation, we do not know whether it was a Varanoid lizard with a very elongated body or a primitive snake. The problem remains unsolved as indicated by its scientific name. Others dating from the middle Cretaceous are snakes like *Pouitella* and *Simohophus* and enigmatic fossils (lizards or snakes?) such as *Mesophus, Pachyphus,* and *Estesius."*

Chapter 18: Lizards

1. *Homeosaurus* (n.d.). Retrieved July 20, 2008, from http://en.wikipedia.org/wiki/Homeosaurus. "Homeosaurus...looks similar to the modern tuatara, to which it is closely related."

Chapter 19: Turtles

1. Dinosaur-era turtle, *Plesiobaena antiqua,* from Hell Creek Formation, on display at the Milwaukee Public Museum, species identified by Dr. Peter Sheehan, head of Geology Department at the Milwaukee Public Museum, adjunct professor, Department of Geosciences at the University of Wisconsin, Milwaukee.

2. Modern pond turtle, *Trachemys scripta,* species identified by Paul Cook and Mr. Jon Wiebe, Louisiana Department of Wildlife and Fisheries.

3. *Turtle* (n.d.). Retrieved July 26, 2008, from http://en.wikipedia.org/wiki/Turtle. "Turtles are divided into three suborders, one of which, the Paracryptodira, is extinct. The two extant suborders are the Cryptodira and the Pleurodira. The Cryptodira is the larger of the two groups and includes all the marine turtles, the terrestrial tortoises, and many of the freshwater turtles. The Pleurodira are sometimes known as the side-necked turtles, a reference to the way they withdraw their heads into their shells. This smaller group consists primarily of various freshwater turtles."

4. Kazlev, M. A., *Norian Age - 3, Tetrapods of the early Late Norian (Sevatian) -Age of Sauropodomorphs* (2005). Retrieved March 1, 2008, from Palaeos.com/Mesozoic/Triassic/Norian.3.htm. "Proterochersis robusta is a more advanced form, the earliest known of the side-necked turtles or pleuodires."

Author's Note: Norian Age is Triassic.

5. Joyce, Walter G. (July 2000). The First Complete Skeleton of Solnhofia parsoni (Crytodira, Eurysternidae) from the upper Jurassic of Germany and its taxonomic implications. *Journal of Paleontology, Volume 74, Issue 4,* pp. 684–700.

Author's Note: Article says that the Jurassic Plesiochelys is a Cryptodira: "A complete skeleton of Solnhofia parsonsi (Cryptodira, Eurysternidae) from the Kimmeridgian/Tithonian boundary of Schamhaupten, Germany, provides the first complete understanding of the postcranial morphology of this genus. The here newly described postcranial characters are important in distinguishing Solnhofia from shell-based genera and thus help in resolving part of the parataxonomic conflict between shell-based and cranium-based turtle genera. This disparity originated during the last 150 years due to the history of fossil finds, preparation, and changing interests of researchers. Synonymies of Solnhofia with such turtle genera as Eurysternum, Idiochelys, Plesiochelys, Thalassemys, and Euryaspis can now be refuted. Similarities with Hydropelta are apparent, but not considered sufficient to support a synonymy. Newly observed or confirmed characters include the relatively large head (40 percent of the carapace length), the pentagonal carapace, the unique arrangement of bones and fontanelles in the pygal region, and the absence of mesoplastra, epiplastra, and an entoplastron."

6. *Australian Snake-necked Turtle* (n.d.). Retrieved July 26, 2008, from http://www.turtlepuddle.org/exotics/chelodina.html. "Called the Australian snake-necked turtle, common snake-necked turtle, or Eastern Long Neck Tortoise, its scientific name is *Chelodina longicollis*. This side-necked turtle is one of several species that is characterized by its extremely long neck, almost as long as the rest of its body."

7. *Terrapene carolina* (n.d.). Retrieved July 26, 2008, from http://species.wikimedia.org/wiki/Terrapene_carolina.

Author's Note: Article classifies this box turtle in the suborder of Cryptodira.

8. *Trachemys* (n.d.). Retrieved July 26, 2008, from http://species.wikimedia.org/wiki/Trachemys.

Author's Note: Article classifies this living pond turtle in the suborder of Cryptodira.

Chapter 20: Birds

1. Stidham, T., *Nature*, A lower jaw from a Cretaceous parrot, November 5, 1998, Volume 396, pp. 29-30.

2. Interview with Dr. William Clemens, professor of integrated biology, curator of the Museum of Paleontology, University of California, Berkeley, conducted

November 9, 1998, by author for video series *Evolution: The Grand Experiment*.

3. Strickberger, M. (1996). *Evolution* (2nd Edition). Boston: Jones and Bartlett Publishers. Page 409.

4. Interview with Dr. Paul Sereno, paleontologist and professor at the University of Chicago, conducted February 24, 1999, by author for video series *Evolution: The Grand Experiment*.

5. Sanders, R., (Nov. 1998), *Berkeleyan,* A newspaper for faculty and staff at the University of California, Berkeley, "Parrot Fossil from the Cretaceous Pushes Back Origin of Modern Land Birds." "Until now the only modern bird fossils uncovered from the Cretaceous have been water birds: loons, duck-like waterfowl, shorebirds and tube-nosed seabirds like the albatross. The oldest of these dates from about 80 million years ago."

Chapter 21: Mammals

1. Interview with Dr. William Clemens, professor of integrated biology, curator of the Museum of Paleontology, University of California, Berkeley, conducted November 9, 1998, by author for video series *Evolution: The Grand Experiment*.

2. Rowe, T., "Roots of the Mammalian Family Tree", March 25, 1999. Article accessed May 10, 2004 from http://www.carnegiemnh.org/news/99-jan-mar/990322jeholodens_comments.html. "The first Mesozoic mammals were discovered in 1812 by a mason in a tilestone quarry near Heddington, England. These fossils came from the Middle Jurassic (roughly 165 million year old) Stonesfield Slate, and consisted of two isolated lower jaws, each belonging to a different species. Today the Stonesfield jaws are still the oldest-known fossils of the 'crown clade' Mammalia...."

3. *Morrison Formation* (Last Modified August 10, 2008). Retrieved August 10, 2008, from http://en.wikipedia.org/wiki/Morrison_Formation.

Author's Note: Article lists these 43 species of mammals found in this one rock formation:

Genus Docodon (5): *D. victor, D. striatus, D. affinis, D. crassus, D. superus.*

Genus Fruitafossor (1): *F. windscheffeli.*

Genus Ctenacodon (4): *C. serratus, C. laticeps, C. nanus, C.scindens.*

Genus Psalodon (3): *P. potens, P. fortis, P. marshi.*

Genus Zofiabaatar (1): *Z. pulcher.*

Genus Glirodon (1): *G. grandis.*

Genus Priacodon (4): *P. robustus, P. gradaevus, P. lulli, P. fruitaensis.*

Genus Trioracodon (1): *T. bisulcus.*

Genus Aploconodon (1): *A. comoensis.*

Genus Comodon (1): *C. gidleyi.*

Genus Triconolestes (1): *T. curvicuspis.*

Genus Amphidon (1): *A. superstes.*

Genus Tinodon (2): *T. bellus, T. lepidus.*

Genus Eurylambia (1): *E. aequicrurius.*

Genus Paurodon (1): *P. valens.*

Genus Tathiodon (1): *T. agilis.*

Genus Archaeotrigon (2): *A. brevimaxillus, A. distagmus.*

Genus Araeodon (1): *A. intermissus.*

Genus Foxraptor (1): *F. atrox.*

Genus Euthlastus (1): *E. cordiformis.*

Genus Pelicopsis (1): *P. dubius.*

Genus Comotherium (1): *C. richi.*

Genus Dryolestes (3): *D. priscus, D. tenax, D. obtusus.*

Genus Laolestes (2): *L. eminens, L. grandis.*

Genus Amblotherium (1): *A. gracilis.*

Genus Miccylotyrans (1): *M. minimus.*

4. Interview with Dr. Zhe-Xi Luo, curator of vertebrae paleontology and associate director of research and collections at the Carnegie Museum of Natural History, conducted May 17, 2004, by author for video series *Evolution: The Grand Experiment*. Question: "How many mammal fossils have been found in the dinosaur layers?" Dr. Luo: "There are about 280 different groups or genera of fossil mammals and of this, the vast majority are just fragments."

5. Museum display "Age of Reptiles" photographed at Sam Noble Oklahoma Museum of Natural History (April 2002).

6. Interview with Dr. Zhe-Xi Luo, curator of vertebrae paleontology and associate director of research and collections at the Carnegie Museum of Natural History, conducted May 17, 2004, by author for video series *Evolution: The Grand Experiment*.

7. Interview with Dr. Donald Burge, curator of vertebrate paleontology, College of Eastern Utah Prehistoric Museum, conducted February 13, 2001, by author for video series *Evolution: The Grand Experiment*.

8. Interview with Dr. Thomas Rich, curator of verte-

brate paleontology at Museum Victoria in Melbourne, Australia, conducted March 3, 2005, by author for video series *Evolution: The Grand Experiment*.

9. Duda, Kathryn, associate editor of *Carnegie* Magazine (1997). *The Age of Dinosaurs Lives On.* Retrieved March 4, 2008, from http://www.carnegiemuseums.org/cmag/bk_issue/1997/julaug/feat2.htm. "The largest known Mesozoic mammal was *Gobiconodon*, which weighed 10–12 pounds, measured 18-20 inches and **might have resembled a large opossum,** but more robust and powerful, and less agile. It was likely more of a walker than a runner and climber, and fossil sites show that it coexisted with several kinds of dinosaurs. Its sharp teeth confirm that it was a predator or scavenger."

10. *Jurassic Landscape, Gobiconodon detail.* (n.d.). Retrieved from the Karen Carr Studio, Inc., website, www.karencarr.com/tmpl1.php?CID=128, on March 4, 2008. Ms. Carr is the scientific illustrator for the Sam Noble Oklahoma Museum of Science and Natural History and her painting of this mammal is on display at this museum. She describes *Gobiconodon* as an "early **possum-like animal.**"

11. Duda, Kathryn, *The Age of Dinosaurs Lives On* Accessed September 8, 2008, from http://www.carnegiemuseums.org/cmag/bk_issue/1997/julaug/feat2.htm

12. Interview with Dr. Zhe-Xi Luo, curator of vertebrate paleontology and associate director of research and collections at the Carnegie Museum of Natural History, for video series *Evolution: The Grand Experiment*, conducted May 17, 2004, by author. "So in addition to the teeth, we have the ankle bones, and we have the wrist bones to help to establish either *Eomaia Eomania* to be placed on placental or *Sinodelphys* to be placed on the marsupial line."

Author's Note: Both *Eomania* and *Sinodelphys* are dinosaur age mammals.

13. Sorin, A. and P. Myers. 2000. "Monotremata" (On-line), Animal Diversity Web. Accessed March 4, 2008, at http://animaldiversity.ummz.umich.edu/site/accounts/information/Monotremata.html. "Monotremes are restricted to Australia and New Guinea. Their fossil record is very poor; the earliest fossil attributed to this group is from the early Cretaceous. A fossil from Argentina suggests that the monotremes were more widely distributed early in their history."

14. American Association for the Advancement of Science News Release, *Jurassic "Beaver" From China,* February 24, 2006, accessed September 9, 2008, from www.aaas.org/news/releases/2006/0224beaver.shtml

15. World Science, *Jurassic "Beaver" From China Found*, February 27, 2006, accessed September 9, 2008, from www.world-science.net/othernews/othernews-nfrm/060227_beaver.htm

16. Lemonick, M., *Time*, January 24, 2005, p. 56. "It had a robust body, with short legs that splay out to the side, similar to a Tasmanian devil."

Chapter 22: Cone-Bearing Plants

1. Guy Darrough, president of Lost World Studios, St. Louis, MO a (self-taught) paleontologist. His dinosaur models and fossils are on display in various museums, botanical gardens, and zoological parks. Interview conducted with Mr. Darrough December 30, 2003, by author for video series *Evolution: The Grand Experiment.* "A number of living fossils that are still around today that are **almost identical** to the ones that lived during the Mesozoic would be the ginkgo, the metasequoia, horsetail rushes, even a common little water plant called the mosquito fern."

2. Dinosaur-era ginkgo from the Hell Creek Formation on display at the Milwaukee Public Museum, *Ginkgo adiantoides,* identified by Dr. Peter Sheehan, head of Geology Department at the Milwaukee Public Museum, adjunct professor, Department of Geosciences at the University of Wisconsin, Milwaukee.

Chapter 23: Spore-Forming Plants

1. Dinosaur-era fern, *Cladophlebis australis,* on display at Museum Victoria, Melbourne, species identified by David Pickering, collections manager, Museum Victoria.

2. Living tree fern, *Cyathea medullaris*, identified by Rainforestation Nature Park duty manager, Australian Rain Forest, Cairns, Australia.

3. Living giant Australian Tree Fern, *Cyathea cooperi* identified by Rainforestation Nature Park duty manager, Australian Rain Forest, Cairns, Australia.

4. Dinosaur-era Sensitive Fern from the Hell Creek Formation on display at the Milwaukee Public Museum, *Hydropteris pinnata,* identified by Dr. Peter Sheehan, head of Geology Department at the Milwaukee Public Museum, adjunct professor, Department of Geosciences at the University of Wisconsin, Milwaukee.

5. See Chapter 22, footnote 1.

6. Dinosaur-era fern from the Petrified Forest National Park, *Cynepteris lasiophora,* identified by Dr. Sid Ash, adjunct professor, Department of Earth and Planetary Sciences, University of New Mexico, Albuquerque, NM.

7. Office of Public Works, National Botanic Gardens Glasnevin Home Page, *The Flora of Ireland,* accessed March 24, 2008, www.botanicgardens.ie/herb/census/flora.htm. "There are three major types of bog in Ireland: firstly Fens, which form where the bog is fed from ground waters rich in nutrients; raised bogs, which occupy the sites of former lake basins, and often form on top of fens, especially in the Irish midlands; and lastly blanket bogs, which cover mountain tops or sloping ground, especially on the west coast....**Raised bog is almost pure sphagnum moss** with scattered grasses and sedges, and these are the bogs from which moss peat is extracted."

Chapter 24: Flowering Plants

1. Druyan, A. (executive producer), and Gibson, K. (producer) (2000). *Cosmos. Carl Sagan. Episode II: One Voice in the Cosmic Fugue* (DVD Video - 7 Disc Collector's Edition). (Available from Cosmos Studios, Inc., 11440 Ventura Boulevard, Suite 200, Studio City, CA 91604-3145.) In this episode, Carl Sagan states: "But the dinosaurs still dominated the planet. Then suddenly, without warning, all over the planet at once the dinosaurs died. The cause is unknown but the lesson is clear. Even 160 million years on a planet is no guarantee of survival. **The dinosaurs perished around the time of the first flower.**"

2. Levin, H. (1996). *The Earth through Time.* (5th Edition). Fort Worth: Saunders College Publishing. p. 432

3. Cornet, Bruce (last updated 2008). *Upper Cretaceous Facies, Fossil Plants, Amber, Insects and Dinosaur Bones, Sayreville, New Jersey.* Retrieved March 21, 2008, from http://www.sunstar-solutions.com/sunstar/Sayreville/Kfacies.htm#Facies%20C. "Facies C Plant Fossils. This facies contains the greatest diversity of angiosperm leaves found in the quarry. There are Platanaceous leaves (Plane tree family - includes *Platanus*), Ericaceous leaves (Heather family - includes *Rhododendron, Kalmia*), possible Lauraceous leaves (Laurel Family), and grass-like leaves (Poales; formerly Gramineae). Many of these plants may have grown on delta levees and overbank deposits, close to the site of deposition."

4. From museum display (August 2008) at the Carnegie Museum of Natural History, Pittsburgh, PA. "The CMNH [Carnegie Museum of Natural History] oviraptorosaur inhabited a humid, forested plain that was populated by a rich diversity of plants and animals. This environment is preserved in a rock unit that is known today as the Hell Creek Formation. Flowering plants, which had evolved earlier in the Cretaceous Period, were diverse and abundant in this ecosystem. Plants represented in our exhibit include *Ginkgo adiantoides* (similar to the modern ginkgo tree), *Sabalites* (a fan palm), *Platanus raynoldsii* (a relative of modern sycamore trees), and *Palaeoaster inquirenda* (a relative of modern poppies)."

5. Dinosaur-era lily pad from the Hell Creek Formation on display at the Milwaukee Public Museum, *Nelumbium montanum,* identified by Dr. Peter Sheehan, head of Geology Department at the Milwaukee Public Museum, adjunct professor, Department of Geosciences at the University of Wisconsin, Milwaukee.

6. Living poplar leaf, *Populus szechuanica,* identified by Pierre-yves Landover, Les.arbres.free.fr.

7. Living ash leaf identified by Joel Reynaud, Laboratoire de Botanique, Lyon, France, as *Fraxinus excelsior*.

8. Modern soapberry leaf, *Sapindus saponaria,* identified by John Seiler, Department of Forestry, Virginia Tech.

9. Modern California bay leaf, *Umbellularia californica,* identified by David Magney, Environmental Consulting.

10. Interview with Dr. Peter Crane, director of the Royal Botanic Gardens in London, England, conducted by author August 26, 2002, for video series *Evolution: The Grand Experiment*. Professor Sir Peter Crane is one of the world's leading experts in plant evolution. Dr. Crane holds academic appointments in the Department of Botany at the University of Reading and the Department of Geology at the Royal Holloway College.

11. Modern magnolia, *Magnolia yunnanensis,* identified by Dennis Werner, JC Raulston Arboretum, North Carolina State University.

12. West Indian thatch palm, *Thrinax radiata,* identified by Anne Muecke, Horticopia.com.

13. Museum display photographed at the New Mexico Museum of Natural History and Science in Albuquerque,

(March 19, 2002).

14. Indian laurel, *Ficus nitida* , identified by Walter Warriner, consulting arborist, Redondo Beach, California.

15. Madagascar palm, *Pachipodium lamerii,* identified by Dr. Svetlana Belorustseva, Moscow State University, Russia.

16. Display sign photographed at the Milwaukee Public Museum, Wisconsin (February 7, 2002).

Chapter 25: Coming Full Circle - My Conclusions

1. Interview with Dr. John Long, head of science at Museum Victoria, Melbourne, Australia, for video series *Evolution: The Grand Experiment,* conducted March 8, 2005, by author. Dr. John Long estimates the total number of fossils in museums worldwide as "hundreds of millions." Dr. Long: "....And now, in any museum around the world, you can see millions of fossils. This museum alone has four million fossils in its paleontology collection, and we're only one museum in Australia. So, if you went around the whole world and you added up how many fossils there are, you'd get hundreds of millions of fossils; and this is the modern overwhelming evidence for evolution."

2. Ramel, Gordon (Updated May 2008). *Welcome To the Myriad Worlds of the Invertebrates.* Retrieved March 25, 2008, from http://www.earthlife.net/inverts/an-phyla.html.

Author's Note: Animals are divided into 34 phyla. I have arbitrarily assigned the most recognizable seven phyla as "major" and the more obscure or diminutive 27 phyla as "minor." Other authors may choose different phyla for these two divisions.

Major Living Animal Phyla

Scientific Name	Examples
Chordata	Fish, Amphibians, Reptiles, Mammals, Birds
Echinodermata	Starfish, Brittle Stars, Sea Urchins, Sea Cucumbers
Arthropoda	Insects, Crabs, Spiders, etc.
Mollusca	Oysters, Snails, Nautilus
Cnidaria	Sea Anenomes, Jellyfishes
Porifera	Sponges
Annelida	Earthworms, Tube Worms

Minor Living Animal Phyla

Scientific Name	Examples
Pogonophora	Beard Worms
Placozoa	Placozoa
Chaetognatha	Arrow-worms
Mesozoa	Mesozoa
Nemertina	Ribbon Worms
Gnathostomulida	Sand Worms
Priapulida	Phallus Worms
Sipuncula	Peanut Worms
Echiura	Spoon Worms
Gastrotricha	Gastrotrichs
Rotifera	Rotifers
Platyhelminthes	Flat Worms
Tardigrada	Water Bears
Pentastoma	Tongue Worms
Onychophora	Peripatus
Nematomorpha	Horsehair Worms
Kinorhyncha	Spiny-crown Worms
Loricifera	Brush Heads
Acanthocephala	Spiny-headed Worms
Nematoda	Roundworms
Cycliophora	Cycliophorans
Entoprocta	Marine Mats
Ectoprocta	Bryozoans
Phoronida	Phoronans
Ctenophora	Comb Jellies
Brachiopoda	Lamp Shells
Hemichordata	Hemichordates

3. *Plants* (n.d.). Retrieved March 27, 2008, from http://en.wikipedia.org/wiki/PlantPlant.

Author's Note: Of the 12 divisions of plants, I consider five of these to be of minor divisions because of their diminutive size or obscure nature, namely Chlorophyta and Charophyta (the algaes), Marchantiophyta (liverworts), Anthocerotophyta (hornworts), and Gnetophyta (gentophytes). I do not mean to imply that these fossils do not exist in dinosaur fossil layers. Also, it should be noted that different scientists classify plants differently. Some consider one group a division while others would not, etc. In each scenario, scientists would come up with a different total number of divisions.

Major Living Plant Divisions

Group	Division name	Common name
Seed plants	Magnoliophyta	Flowering plants

Seed plants	Pinophyta	Conifers
Pteridophytes	Pteridophyta	Ferns, horsetails
Seed plants	Cycadophyta	Cycads
Seed plants	Ginkgophyta	Ginkgo
Bryophytes	Bryophyta	Mosses
Pteridophytes	Lycopodiophyta	Club mosses

Minor Living Plant Divisions

Group	Division name	Common name
Green algae	Chlorophyta	Green algae
Green algae	Charophyta	Green algae
Bryophytes	Marchantiophyta	Liverworts
Bryophytes	Anthocerotophyta	Hornworts
Seed plants	Gnetophyta	Gnetophytes

4. See Chapter 1, footnote #2.

5. Lerner, Eric. (2004). *New Scientist,* Bucking the Big Bang, May 22, 2004, p. 2448.

6. Milner, R. (1990). *The Encyclopedia of Evolution: Humanity's Search for Its Origins.* New York: Facts on File Publishers. pp. 3-4.

7. Interview with Dr. Andrew Knoll, professor of biology, Harvard University, conducted October 13, 1998, by author for video series *Evolution: The Grand Experiment*. Dr. Knoll: "Darwin devotes two chapters of *The Origin* [*of Species*] to the fossil record. And you might think that's because Darwin, like most of his intellectual descendants, would have seen the fossil record as the confirmation of his theory. That you could really see, directly document, the evolution of life from the Cambrian to the present. But, in fact, when you read *The Origin* [*of Species*], it turns out that Darwin's two chapters are a carefully worded apology in which he argues that natural selection is correct despite the fact that the fossils don't particularly support it."

8. *Turtles* (n.d.). Retrieved July 20, 2008, from http://en.wikipedia.org/wiki/Turtle. "The first turtles are believed to have existed in the early Triassic Period of the Mesozoic era, about 200 million years ago. Their exact ancestry is disputed...The earliest known turtle is proganochelys, though this species already had many advanced turtle traits, and thus probably had many millions of years of preceding turtle evolution and species in its ancestry. It did lack the ability to pull its head into its shell (and it had a long neck), and had a long, spiked tail ending in a club, implying an ancestry occupying a similar niche to the ankylosaurs (though only through parallel evolution)."

9. Werner, C. (2007). *Evolution: The Grand Experiment, The Quest for an Answer*. Green Forest: New Leaf Press. pp. 152-153.

10. Werner, C. (2007). *Evolution: The Grand Experiment, The Quest for an Answer*. Green Forest: New Leaf Press. pp. 179-183.

11. Werner, C. (2007). *Evolution: The Grand Experiment, The Quest for an Answer*. Green Forest: New Leaf Press. pp. 174-178.

12. Werner, C. (2007). *Evolution: The Grand Experiment, The Quest for an Answer*. Green Forest: New Leaf Press. pp. 139-143.

Appendices

1. Coelacanth (n.d.). Retrieved March 11, 2008, from http://en.wikipedia.org/wiki/Coelacanth. "Timeline of discoveries. December 23, 1938: Discovery of the first modern coelacanth 30 km SW of East London, South Africa."

2. Display photographed at the Carnegie Museum of Natural History, Pittsburgh, PA (May 17, 2004).

Glossary

1. *Galileo Galilei* (n.d.). Retrieved April 3, 2008, from http://en.wikipedia.org/wiki/Galileo_Galilei. "Galileo's championing of Copernicanism was controversial within his lifetime. The geocentric view had been dominant since the time of Aristotle."

2. Strickberger, M. (1996). *Evolution* (2nd Edition). Boston: Jones and Bartlett Publishers. p. 604. "An existing species whose similarity to ancient ancestral species indicates that very few morphological changes have occurred over a long period of geological time."

Introduction: Dinosaur-era scallop, *Pseudopecten aequivalis*, Jurassic, England, courtesy Dr. Jonathan Radley, Warwickshire Museum, England; Sea star, *Terminaster cancriformis*, Jura Museum, Eichstaat, Germany.

Chapter 1: Embryo drawings from Haeckel's *History of Creation*, Plate III, Volume I, Fourth Edition, D. Appleton and Company, New York, 1893; Brian Greene photo by Alexanian Stock Images, courtesy Nubar Alexanian; Dr. David Gross photo by Rod Rolle.

Chapter 3: Wolf photo taken at Grizzly Discovery Center, West Yellowstone, Montana; dog skeleton photos taken at University of Missouri School of Veterinary Medicine, courtesy of Bobby Colley; live dog photos courtesy of Jupiter Images and Istockphotos; Australia Aborigine photos courtesy Frank Crocker, Rainforestation, Cairns, Australia.

Chapter 4: Title page photo of brittle stars by Ed Bowlby, National Oceanic and Atmospheric Adminiistration (NOAA), taken off coast of Washington, Olympic Coast; live brittle star Sea of Alaska, photo by NOAA; living purple heart sea biscuit photo by Laurie Campbell, Scotland, UK; fossil sea cucumber photo taken by Dr. Richard A. Paselk, Humboldt State University; *Psittacosaurus* dinosaur photo taken at Wyoming Dinosaur Center, Thermopolis; living yellow feather taken by Keoki Stender, Fishpics Hawaii; sauropod dinosaur drawing taken at Sam Noble Oklahoma Museum of Science and Natural History, Norman.

Chapter 5: Live spiny lobster photo by Keoki Stender, Fishpics Hawaii; live crawfish photos taken at JT's Seafood, Lake Charles, Louisiana.

Chapter 6: Modern termite nest, *Coptotermes acinaciformis*, photo by Simon Fearby, public domain; adult mayfly photo by Laurie Campbell, Scotland, UK; living mayfly larvae by Janet and John Garrett, Ipswich, UK; *Ornithodesmus* (pterosaur) painting at Sam Noble Oklahoma Museum of Science and Natural History, Norman.

Chapter 7: Living cockscomb oyster © 2009 Corel.

Chapter 8: *Coelophysis* dinosaur model photographed at Missouri Botanical Gardens, courtesy Living World Studios, St. Louis, Missouri.

Chapter 9: Modern lampshell photo, *Rhynchonella psittacea*, and sea cradle photo, *Stenosemus albus*, by Dr. Svetlana Belorustseva, Moscow State University, Russia; Other sea cradle photo © 2009 Corel.

Chapter 10: Modern tube worm photo by NOAA and College of Charleston.

Chapter 11: Living demosponge photo courtesy of Dr. Svetlana Belorustseva, Moscow State University, Russia; living glass sponge courtesy of Brooke et al, NOAA-CE, HBOI, http://oceanexplorer.noaa.gov/explorations/05deepcorals/logs/hires/fig2cupsponge_hires.jpg; living humpback coral photo by Keoki Stender, Fishpics Hawaii.

Chapter 12: *Torvosaurus* painting by Paul Koroshetz, Mesalands Community College's Dinosaur Museum, Tucumcari, New Mexico; yellowstripe scad photo by Andy Pearson, www.gulliblestravels.co.uk; gissu fish photo by Teruo Okamoto, Saitama, Japan, and translation assistance by Yuko Stender, Fishpics Hawaii; orange roughy and Atlantic herring photos by Thomas Wenneck, Norway; ladyfish photo by Keoki Stender, Fishpics Hawaii; and sardine photo by Johnny Jensen, Denmark, www.jjphoto.dk.

Chapter 13: Pacific angel shark by Bill Barss, Oregon Department of Fish and Wildlife; Shovelnose Ray photo by Robert Perry, Malibu, California, USA.

Chapter 14: Close-up photos of lamprey mouth by David Hansen, University of Minnesota Agricultural Experiment Station; hagfish photo by Peter Batson, Fiordland, New Zealand, ExploreTheAbyss.com.

Chapter 15: Live hellbender photos by Jeffrey Humphries, Clemson University, South Carolina; meat-eating dinosaur photo Carnegie Museum of Natural History; frog painting from Sam Noble Oklahoma Museum of Science and Natural History, Norman.

Chapter 16: Live crocodile photographed at Rainforestation, Cairns, Australia; no swimming sign, Parndana Wildlife Park, Kangaroo Island, Australia; modern gavial skull photo by Tosh Odano of Valley

Anatomical Preparations; live gavial photo provided by Critter Zone, Copyright 2006.

Chapter 17: Live boa constrictor photographed at World Aquarium, St. Louis, Missouri.

Chapter 18: Live gliding lizard photos taken in Malaysia by George Sly, Union High School, Dugger, Indiana.

Chapter 19: Title page photo of eastern box turtle by Critter Zone, Copyright 2006; *Camptosaurus* dinosaur model photographed at Dinosaur National Monument, Vernal, Utah; dinosaur painting on display at Paleontology Museum, Ghost Ranch Conference Center, Abiquiu, New Mexico, by M. Colbert; fossil turtles A, C, H from Harvard Museum of Paleontology, Boston, Massachusetts; fossil turtles D, E, F, G from Royal Belgian Institute of Natural Sciences, Brussels, Belgium; fossil turtle B from Wyoming Dinosaur Center, Thermopolis; live soft-shell turtle photographed in Ste. Genevieve, Missouri.

Chapter 20: Modern avocet standing in water photo by James Ownby; *T. rex* photo taken at Utah Field House of Natural History State Park Museum, Vernal, Utah; *T. rex* painting by Michael Skrepnick displayed at the Carnegie Museum of Natural History; *Pterodactylus* painting by Fred Wilhelm dsplayed at the Geological Museum, University of Wisconsin, Madison.

Chapter 21: Title page hedgehog photo by iStock; *Stegosaurus* model on display at Dinosaur National Monument, Vernal, Utah; age of reptile museum sign on display at Sam Noble Oklahoma Museum of Science and Natural History, Norman; modern European hedgehog, *Erinaceus europaeu*, photo by Laura Campbell, Scotland, UK; modern ringtail opossum photo by Julian Robinson, Australia; Virginia opossum photo courtesy South Carolina Department of Natural Resources; modern tree shrew photo by Dr. Paddy Ryan, Ryanphotographic.com. Thornton, Colorado; on last page of chapter, mammal A, B, D, F, G, and H photographed at Sam Noble Oklahoma Museum of Science and Natural History, Norman; mammal C photographed at Carnegie Museum of Natural History, Pittsburgh, Pennsylvania; and mammal E photographed at California Academy of Sciences, San Francisco.

Chapter 22: Coast redwood cones photo by Mike Clayton, University of Wisconsin, Madison; Bald cypress swamp photo by Bill Lea, USDA Forest Service, Southern Research Station.

Chapter 23: Title page photo of live club moss, *Lycopodium clavatum,* and other live club moss photo, *Lycopodium selago,* by Laurie Campbell, Scotland, UK; dinosaur photo taken at Houston Museum of Natural Science, Texas.

Chapter 24: *Time* Magazine cover courtesy of Getty Images; poplar leaf, *Populus szechuanica,* photo by Pierre-yves Landover, Les.arbres.free.fr; ash leaf, *Fraxinus excelsior,* photo courtesy of Joel Reynaud, Laboratoire de Botanique, Lyon, France; modern soapberry leaf, *Sapindus saponaria,* photo courtesy of John Seiler, Department of Forestry, Virginia Tech; California bay leaf, *Umbellularia californica,* photo by David Magney, Environmental Consulting, Ojai, California; magnolia, *Magnolia yunnanensis*, photo by Dennis Werner, JC Raulston Arboretum, North Carolina State University; thatch palm photo courtesy of Anne Muecke, Horticopia.com; Indian laurel photo by Walter Warriner, consulting arborist, Redondo Beach, California; Madagascar palm photo courtesy of Dr. Svetlana Belorustseva, Moscow State University, Russia; live plants A, E, F, and G taken at Missouri Botanical Gardens, St. Louis, Missouri.

Chapter 25: Photos in wheel and spreads identified in earlier chapters; modern European hedgehog, *Erinaceus europaeu*, photo by Laura Campbell, Scotland, UK; Flamingo, St. Louis Zoo; shrimp photo by Dr. Svetlana Belorustseva, Moscow State University, Russia;

Appendix B: Photo of live coelocanth fish taken at California Academy of Sciences, San Francisco; fossil coelocanth, fossil lobster, and fossil crab photos taken at University of Wisconsin, Madison, Geological Museum; live sea pen, photo by Ian Skipworth ©

Ianskipworth.com; fossil sea pen photographed at the South Australian Museum, Adelaide; fossil possum skeleton taken at the Mississippi Natural History Museum, Jackson, USA; fossil starfish taken at the Carnegie Museum of Natural History, Pittsburgh, Pennsylvania.

Volume III Advertisement Photos: Photos courtesy of Dr. Clark Howell, University of California, Berkeley.

Other Photos: Photos of Charles Darwin, courtesy of Down House, England; some photos courtesy of National Oceanic and Atmospheric Administration, Washington, D.C.; other photos © 2009 Jupiterimages Corporation, © 2009 iStockphoto, © Photos.com or © 2009 Audio Visual Consultants, Inc. All photos © 2009 Audio Visual Consultants, Inc., were taken by Carl and Debbie Werner.

TEACHER'S MANUAL

Evolution: The Grand Experiment Volume II: Living Fossils Teacher's Manual was designed to teach a critical examination of the theory of evolution to students in public and private schools from grades 7-14.

As one of the most controversial topics of our day, it behooves all of us to review the current evidence both for and against the theory of evolution. The material presented in this course contains interviews with expert scientists from some of the most highly acclaimed scientific institutions, universities, and museums of the world.

This teacher's manual contains all of the tools needed to assist a teacher, including:

- Purpose of Chapter – To summarize the purpose of each chapter
- Class Discussion Questions – To stimulate student interest in the chapter material
- Objectives of Chapter for Students – To assist students in preparing for examinations
- Chapter Examinations – To assess if students adequately understand the materials
- Sectional Examinations – To prepare students for the comprehensive final examination
- Comprehensive Final Examination – To assure students retain the information

The entire teaching program is a turnkey system and does not require scientific expertise to teach the course.

VOLUME III

Human Evolution!

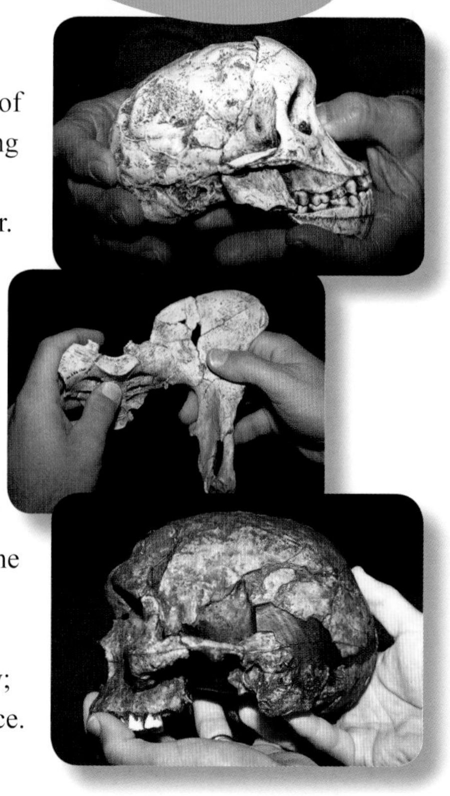

In the third volume of this series, Dr. Carl Werner dissects the complex theory of human evolution in a simple, straightforward style. He accomplishes this using graphs, photos, charts, and interviews with scientists personally involved in the discoveries of *Australopithecus*, Neanderthal, *Homo Erectus* and many more. Dr. Werner, in his 30-year effort to get to the bottom of the theory of evolution, asks if human evolution is real.

Dr. Werner and his wife, Debbie, filmed interviews on location in Australia, England, Germany, and the United States with such notables as Dr. Donald Johanson, the discoverer of Lucy, the late Dr. F. Clark Howell, leader of the international expedition to southern Ethiopia; Dr. Charles Oxnard, recipient of the Charles Darwin Lifetime Achievement Award; Dr. Gert Wenegert, curator of the Neanderthal Museum in Germany; Dr. Ralf Schmitz, archeologist of the Rhine State Department of Archeology; Dr. Angela Milner, Natural History Museum, London; Dr. Daniel Lieberman, professor of biological anthropology at Harvard University; Dr. Taseer Hussain, professor of human anatomy, Howard University; and Dr. Daniel Gasman, professor of history at John J. College of Criminal Justice.